Praise for Howard Frank Mosher's *North Country*

"*North Country* shows how a little can be enough. Lost souls, lonely places and small pleasures alike have their particular dignity when Mr. Mosher shines the light of his sympathy on them." — *New York Times Book Review*

"Part travelogue, part memoir, part meditation, part exploration . . . After this book, I'll always think of [the northern border] as Mosher Country." — *Boston Sunday Globe*

"Exuberant and affirmative." — *Burlington Free Press*

"I'd put Howard Mosher up on the pedestal I keep for Wallace Stegner, Frederick Turner, Edward Hoagland . . . [Mosher is] a marvelous writer." — Frank McCourt, author of *Angela's Ashes*

"Howard Frank Mosher [is] a combination of Ernest Hemingway, Henry David Thoreau and Jim Harrison from Vermont's Northeast Kingdom." — *Los Angeles Times*

"Mr. Mosher has transformed the northern U.S. frontier and the southern coast of Canada into one long and bountiful literary landscape, and in doing so has made me see America more vividly. His wonderful itinerary is bright with anecdote and history and lore and most importantly with affection for his human subjects." — Richard Ford, Pulitzer Prize–winning author of *Independence Day*

"It is hard to think of anyone else who could have chosen to make just this kind of journey, a pilgrimage, really, with such an observant eye and such respect for the natural world and such empathy with humanity, and then, with inimitable craftsmanship, to go home and write about it." — *Rutland* (Vermont) *Herald*

"What Mosher's northern journey is really about is our society's loss of Eden, the garden we were promised when we came here. The garden we've turned into pulp fiction and rocket ranges. The very fact that this brave book can stir up so many thoughts about the predicaments of civilization is surely an indication that it is well worth reading." — *Ottawa Citizen*

"A book that is as handsomely simple and useful as a piece of handcrafted Shaker furniture . . . a moving, ardent, sometimes elegiac tribute." — *Seattle Times*

"Mosher [is] an inquisitive, unpretentious and good-humored guide. [*North Country*] is as true and magical as a compass." — *Pittsburgh Post Gazette*

"Readers need only pick up *North Country* and open it to any of its fifty-five short chapters to find a self-contained gem of fine writing and masterly storytelling filled with people whose characters resonate long after their tales are told." — *Cleveland Plain Dealer*

"Poignant and wistful yet infused with wry humor." — *Vermont Life*

"*North Country* is a classic road book. You could, with confidence, place this book on the shelf next to such American classics as John Steinbeck's *Travels with Charley* and Jonathan Raban's *Old Glory*." — *Detroit Free Press*

"Mosher is a fine reporter with a novelist's eye." — *USA Today*

"*North Country* is a remarkable book . . . It was my great good fortune, page by page, to travel with [Mosher]." — Howard Norman

NORTH COUNTRY

Books by Howard Frank Mosher

Disappearances

Where the Rivers Flow North

Marie Blythe

A Stranger in the Kingdom

Northern Borders

North Country

NORTH COUNTRY

A Personal Journey Through the Borderland

HOWARD FRANK MOSHER

Howard Frank Mosher

A Mariner Book
Houghton Mifflin Company
BOSTON NEW YORK

FOR PHILLIS

Library of Congress Cataloguing-in-Publication Data
Mosher, Howard Frank.
North country : a personal journey through the borderland / Howard Frank Mosher.
p. cm.
ISBN 0-395-83707-3 ISBN 0-395-90139-1 (pbk.)
1. Northern boundary of the United States — Description and travel. 2. Canada — Description and travel. 3. Frontier and pioneer life — United States. 4. Frontier and pioneer life — Canada. 5. Mosher, Howard Frank — Journeys — United States. 6. Mosher, Howard Frank — Journeys — Canada.
I. Title.
F551.M67 1997
917.304'929 — dc21 96-29517 CIP

Printed in the United States of America

QUM 10 9 8 7 6 5 4 3 2 1

Book design by Melodie Wertelet

Parts of this book appeared, in somewhat different form, in "The Sophisticated Traveler" section of *The New York Times Magazine*, March 3, 1996; *The Boston Globe Magazine*, March 17, 1996; and *Country Journal*, June 1986.

Contents

Part Three
BORDER TOWNS

Part Four
THE OUTLAW TRAIL

Part Five
FRESH STARTS

Yes, yes, yes, you're going on a journey. Anyone who wants to take a trip as badly as you do will find a way to take it. But don't be surprised if it turns out to be entirely different from what you expect.

— A gypsy fortuneteller in New York City, summer of 1993

Prologue

On the American side of the border, especially in remote northern frontiers where there's a long tradition of individualism, there's also a deep awareness of how precious an individual's personal liberty is and how one gets and keeps it. What I think you'll find in your North Country outposts is the very strongest sense of independence, on all levels, left in America today. At the same time, I think you'll also discover that in many places the border has created its own zone of independence from the rest of *both* Canada and the U.S.

— Chris Braithwaite, Canadian-born
Vermont journalist and editor

Ever since my grandparents began taking me on weekend trips to the Adirondacks when I was four years old, traveling north has exhilarated me. Later in my boyhood, my father and uncle and I headed north each summer to fish for trout in Quebec's Laurentian Mountains. I can still vividly recall the almost unbearable excitement I felt on those journeys. Long before we reached the border, I had the dramatic sense that we were entering an altogether different land from the tamer Catskills where I lived at the time: a boreal realm of deep evergreen woods, raw-looking papermill towns whose acrid tang you could smell ten miles away, swift dark rivers with hard-to-pronounce Indian names. And then, toward evening, the quickening anticipation of approaching the customs station and, stretching north to the tundra and beyond, Canada.

Next to going there, I loved nothing better than reading about the North. As a boy I read everything I could get my hands on having to do with the North Woods, the northern Great Plains, the Far North. I devoured all of Jack London and Robert W. Service and memorized the Paul Bunyan tales. Between college semesters, I took extended camping trips on my own, drawn north as surely as geese in the spring, searching in the Adirondacks and Laurentians and Maine's vast forests for that transcendent sense of well-being that I experienced in the northern wilderness and nowhere else — least of all in a classroom. I couldn't explain it, but somehow I felt that the North was where I belonged.

So in 1964, after graduating from college and getting married, I moved with my like-minded wife, Phillis, to Vermont's Northeast Kingdom, just a few miles south of the Canadian border, a remote enclave of narrow glacial lakes and north-running rivers, thickly forested granite mountains, half-forgotten hamlets, and high hill farms, home to some of the most fiercely independent Yankee and French Canadian individualists left on the earth.

I have spent the last three decades living here and chronicling the lives and times of these individualists in my novels and short stories; and as unusual as it is these days to live one's entire adult life in a single place, making our home in this last vestige of an earlier Vermont has turned out to be exactly the right choice. Still, for a number of years, as the ever-encroaching development spreading northward from southern and central Vermont began to reach the Kingdom, and as the old horse loggers and hill farmers and moonshiners and whiskey runners vanished like the dying elms on a thousand New England village greens, I'd had a growing desire to search out what remained of the rest of America's northernmost frontiers, to identify the qualities that characterized these regions, and to assess what they might look like twenty, fifty, one hundred years from now.

In the summer of my fiftieth year the desire to make such a trip acquired a certain urgency. It's not that I was experiencing a midlife crisis — I wanted to celebrate turning fifty by having a midlife adventure instead. So at the end of that summer, thinking now or never, with Phillis's encouragement and support I set out in a rented car — our own had one hundred and thirty thousand miles on it — with a single suitcase, half a dozen notebooks and Papermate refills, my hiking boots and fly rod, a couple dozen of my favorite North Country books, and a brand-new Rand McNally road atlas of the United States and Canada.

My itinerary would be loose, though I decided ahead of time that I would deliberately seek out the wildest and most remote country I could find — places like the big woods and lakes of northern Maine, Michigan, and Minnesota, the northernmost Great Plains along our border with Manitoba and Saskatchewan, the breaks and mountains of Montana's Hi-line, the upper Columbia River Valley, and the Cascades. Having lived hard by the Canadian border for thirty years, I would use it as a compass bearing, venturing into Canada from time to time if I felt like it. For the most part I'd follow my nose and see where it led me.

What did I discover? For starters, just as my friend and neighbor Chris Braithwaite had predicted, I found not so much a border, in the conventional sense, as a vast and little-known territory so distinct from the rest of the United States that it has a special name of its own. From coast to coast it's known as the North Country: an immense, off-the-beaten-track sector of America inhabited by remarkably versatile, resilient, and, most of all, independent-minded people, most of whom are still intimately in touch with the land they live on and with their respective communities.

At the same time, as I traveled westward, impelled by my search for northern lore and history and our last northern frontiersmen and women, I could not seem to escape my own history, so much of which has been informed by my love of all things northern.

The following account of my journey, then, is also a personal memoir: the story of my own life in the North Country and how I came to be a writer here.

Part One

THE GREAT NORTH WOODS

Notes from Route 2

In the old days Hortons Bay was a lumbering town. No one who lived in it was out of the sound of the big saws in the mill by the lake. Then one year there were no more logs to make lumber.

— Ernest Hemingway, "The End of Something"

5:30 A.M. Irasburg, Vermont. I strike off from my home in the Northeast Kingdom on a clear dawn in late August, which also happens to be the morning of the first hard frost of the year. Caught off guard, as usual, I left my car outside last night, so I have to scrape the windows all the way around. Frost, with September still a week away! Yet, so far from discouraging me, the early freeze-up — it's severe enough to kill our tomato plants — seems fitting, emblematic of the harsh territory I'm about to visit.

6:00 A.M. St. Johnsbury, Vermont. I pick up U.S. Route 2 thirty miles south of my home and head due east, into the rising sun, toward New Hampshire and Maine. Route 2 stretches most of the way across the North Country to the Pacific, and during the next six weeks I'll bump into it here and there from Maine to Michigan to Montana, like an old friend. For the most part I'll use it as a kind of unofficial southern boundary; in general I'll stay much closer to the Canadian border. Still, in New England, Route 2 passes through quintessential North Country, including many of the same deep woods and big rivers my father and uncle and I passed on those early fishing trips to Canada. And glory be, I'm nearly as excited now, forty years later, at the beginning of my cross-country odyssey, as I was then, though at the same time I can't shake the feeling that here in the North Woods of New England, I'm witnessing — like Nick Adams in the Hemingway story — the end of something.

* * *

7:00 A.M. Near the New Hampshire state line. For one thing, it's the end of summer, always a reflective time for me. Look at the roadside flowers: tall pink fireweed (the ubiquitous late-summer flower of North Woods clearings from Maine to Washington), goldenrod, asters, and here and there in swampy patches back from the road a soft maple turned red as fire for the fall just ahead. The farmhouses are weathered a November gray, gray as the granite outcroppings in the surrounding fields. The barn roofs, still sparkling with the first frost of the year, have a wintery look, steeply pitched to shed the two to three hundred inches of snow that falls annually in these mountains. The long, linked chains of sheds known as "North Country ells" connect house and barn so that on January mornings when the temperature plummets to forty below zero, farmers can walk from kitchen to milking parlor without stepping outside. Many of these places are abandoned now, the fields behind them overrun with wild redtop grass and alders and clumps of poplars and birches. The self-sufficient family farm has become an anachronism in northern New England just during the thirty years I've lived here. When at last I spot a small herd of Jersey cows headed from barn to pasture, I can't help feeling that this last working dairy farm for ten miles in either direction, where there were once thirty or forty, represents the end of a tradition and a way of life.

8:30 A.M. Up into a steep cut through New Hampshire's Presidential Range, past black-water bogs fringed with jagged necklaces of ice, past a sequestered backwater broken by the ring of a single rising brook trout (reminding me how drastically the brook trout fishing has fallen off in such places over the past three decades), into a region of vast clear-cuts: totally denuded hillsides that look as though they've been subjected to weeks of saturation bombing. I stop briefly a few miles farther along to watch a logging operation in progress. The early-morning woods is filled with the savage whining of two leviathan red-and-orange tree-harvesting machines clipping sixty-foot-tall spruces and firs off at the base like giant flowers. From the treeless bank of a nearby stream comes the ungodly blast of a chipper, louder than a 747's engines at full throttle. Its whirling knives are reducing full-grown trees to hundreds of thousands of fragments the size of your thumbnail in sixty seconds flat, spitting them out in a steady river into the backs of eighteen-wheelers to be rushed to generating stations to fire boilers to manufacture electricity to power, among other things, the colossal papermaking machines, as long as football fields, at Groveton and Rumford and Skowhegan, which convert these wood chips, millions upon millions of them per day, into newsprint and paper towels, toilet tissue and Kleenexes, schoolbooks and, very possibly, the stationery for the rejection letter I received from *Harper's* two weeks ago informing me that a recent story I've written (set in this very countryside) is "too linear and old-fashioned, with its traditional beginning, middle, and end" to be published in their magazine — causing me to reflect, as I nailed the note to the side of my own weathered and disused barn at home and blasted the living hell out of it with my shotgun, what a sorry end *this* pellet-riddled scrap of paper was for a tree that may once have shaded a trout brook or a deer run.

10:00 A.M. Maine, somewhere between Rumford and Skowhegan. Though the temperature is close to seventy and there isn't a cloud in the sky, I can't seem to get away from winter. More than a few of the farmhouses still sport brown wreaths on their doors, left over from last Christmas. RFD mailboxes are mounted on long wooden planks uptilted like seesaws or swing from chains attached to yardarms six feet off the ground, to permit snowplow blades to pass beneath; and a state road crew is already out stringing up snow fencing. Listing, bullet-pocked drive-in theater screens no technicolor feature presentation or titillating coming attractions have flashed across for years, or ever will again, slump into lots overrun with Canadian bull thistles and steeplebush and yellow-flowering mullein plants five feet tall; but what are the coming attractions for these half-forgotten, out-of-the-way mill towns I'm driving through, with their semiabandoned main streets running quickly into the interchangeable edge-of-town commercial strips that the boarded-up downtown stores have defected to? *Is* there any future for small towns in America's North Country, or are they, too, doomed to go the sad way of the last, endangered small farms and big woods? I'll look into that, too, over the following weeks and yes, if necessary, celebrate the diminution of our northern frontier. For above all I'm determined to make this intensely personal journey one of exuberance and affirmation rather than lament and nostalgia. So hurrah for the disused boxcars sitting doorless on grassy sidings off defunct spur lines under bone-dry wooden water towers baking in the hot late-summer sunshine, hurrah for the thirdhand trailers slouched in clearings cluttered with junk and battered pickups, hurrah for the tiny emptying-out mill towns and abandoned farms throughout this vast North Country that I'm about to celebrate, in celebration of turning fifty.

Jumping Off from Lubec

A sardine, me b'y, is a juvenile herring, no more, no less. I've netted tens of millions of 'em in me time, but these days the bottom's fallen right out of the sardine market. Who wants to open a can of oily old sardines when they can get a Big Mac anytime they wants?

— Elisha the double-ender, Lubec, Maine

In the middle of the afternoon, Lubec loomed suddenly out of the blueberry barrens ahead of me, a cluster of houses and abandoned sardine factories shouldered together on a lofty rock promontory jutting into Passamaquoddy Bay, with the Atlantic Ocean just beyond. Perched on its adamantine cape, the town reminded me of the remote Newfoundland fishing outport where my wife and I spent a summer during the late 1960s — a town right out of E. Annie Proulx's novel *The Shipping News.*

I parked my car in front of an empty warehouse next to the American customs station at the west end of the bridge connecting Lubec with Campobello Island, New Brunswick, and started down along the waterfront — a little unsure of how to proceed but eager to see what I could see. To my immediate right, the hulking, empty sardine factories jutted out into the bay on rotting piles a good twenty-five feet above the brackish low tide. To my left the town clung precariously to the steep side of the point. From the waterfront, the crowded houses, nondescript sheds, and tiny stores occupying one or two rooms of people's homes reminded me more than ever of a Newfoundland outport, though the little stores in Lubec today all sell videos as well as beer, cigarettes, and a few staples, and cold cuts instead of cod tongues.

It was out of one of these establishments that the old man materialized, abruptly and apparition-like, with a black watch cap on his head, a six-pack of Budweiser under his arm, and a grim expression on his face.

"Beautiful day," I ventured.

"I'd call her a weather breeder," he said in the tone of a man voicing strong disagreement.

"Weather breeder?"

"Aye. A sunny, dead-calm day before a hell of a blow." He squinted out over the bay and nodded with satisfaction. "She's a-coming, all right. Us'll get a big storm by morning at the latest."

As we fell into step, I introduced myself. "Name's Elisha," my friend said, thrusting a beer into my hand. "Around here, they calls me Old Elisha. Old Elisha the double-ender."

I looked at him inquiringly.

"Spawned of Canadian parents in the U.S. of A.," he explained. "Making me a dual citizen. See this rock?"

Elisha had shifted gears so fast that it took me a moment to realize that he meant the cape that the town had been built on.

"Well, this rock just keeps on a-going down, down, down beneath the sea. You can step right off the edge of town into two hundred feet of freezing salt water, and that's all, she wrote, down you goes to Davy Jones. Point being, in Lubec there's nowhere for the incoming tide to go but up. Swoosh! In she rushes like a steam locomotive coming at you full throttle. In my time, I've seen the tide here run thirty-one feet high."

Elisha's voice throbbed with grisly rapture, as though, with luck, Armageddon itself would begin right in Lubec with a thirty-one-foot tide.

When I told him that Lubec was the jumping-off town for my cross-country trip, he said that at one time it had been a jumping-off town for the whole world. Lumber, cod, halibut, haddock, sardines, salmon, blueberries, and potatoes went out of the harbor on three-masted schooners for ports around the globe. There was plenty of work here, too. As recently as the Second World War, Lubec had ten sardine sheds working round the clock. But like everywhere else, times here had changed. These days you couldn't scare up a halibut or a haddock in the nearby coastal waters to save your life; the cod had been fished nearly out of existence; even lobsters weren't as plentiful as they'd once been. Sardines were still available in good numbers, but the bottom had fallen out of the market.

"Fact is," Elisha said, "in me own lifetime I've seen Lubec go from a prosperous seafaring port to a dying little fishing village. I hopes for the best, me b'y. But I expects the worst, and I've rarely been disappointed."

As we stood together on the dock, watching the tide turn and sipping our beers, I asked Elisha what made Lubec different from any other small town in Maine — a coastal village down around Portland, for instance. "That's easy," he said. "It's the people."

"What's different about them?"

"Stubborn-minded!" he replied with delight. "We wants to keep our own high school even though they's only ninety kids left in town to go to school there. We wants our own waste management system, which is to say dump. Above all we wants to work for ourselves. Pick blueberries out on the barrens, cut a little firewood, pull a few lobster traps. Anything to work for ourselves. We're stubborn, b'y! Us are the stubbornest people on the face of the earth, which we've had to be to survive at all. And unless the high tide washes Lubec right out to sea some fine morning, I expect we'll be right here hanging onto this old rock and being stubborn for a long time to come."

The onset of fall has always stirred deep childhood memories in my mind, and traveling north along the Maine coast and the Canadian border on U.S. 1, late in the evening of this first fall-like day of the year and the first day of my trip, I found myself thinking of the fall day when I was ten years old and about to cross an important border in my life. I was walking up the steep hill of the tiny upstate New York village of Chichester, near where I had been born and grown up, to say goodbye to my uncle and aunt, Reg and Elsie Bennett, because like Lubec in the 1990s, Chichester was dying, and my family was moving away. The local furniture factory had closed for the last time, and there was no town nearby with a school for my father and uncle, both teachers, to work at or for me to attend.

Tucked off in the northern Catskills, near Rip Van Winkle territory, Chichester was a typical North Woods village that could as well have been situated two hundred miles farther upstate on the Canadian border. From the early 1800s until the Depression, it was a company woodworking town. Then in the late 1920s the company began to fail, and in 1939, three years before I was born, the entire town, including the furniture factory, houses, store, spur railroad, and thousands of acres of surrounding timberland, was parceled out and sold at a public auction in a single day. Since then there had been two or three attempts to reestablish the furniture business, but recently the factory had closed again, this time for good.

Just across the deserted street was Chichester's empty company store, its Salada Tea sign aslant in the window, gathering dust. Beyond the store I passed one abandoned house after another, the beautiful scrollwork under the eaves rotting away, and only a faint hint here and there, in sheltered corners, of the gold paint that had once made the whole town glow in the sunshine like a northern El Dorado. I passed the empty house of Chichester's beekeeper, Oat Morse, whose bees, it was said, always swarmed on his chimney to tell him when a friend or neighbor had died. Next door was the sagging, abandoned home of Old Grammy Moon, the town witch, who could keep a candle stub burning all night long and could make a can of yellow-eye beans walk right off her table just by frowning at it. Then the empty place of my friend Lennie Miller, with whom I had combed every square foot of the surrounding mountains for ginseng, blackberries, brook trout, and fiddlehead ferns.

Just up the hill from the Millers' place was the village school where my uncle Reg Bennett had begun teaching when he was sixteen, and the mountain lane that my father and uncle and I would drive up in Dad's old Ford to hear the Red Sox–Yankees games on the car radio. And, just beyond the lane, my aunt and uncle's place.

Uncle Reg came out on the porch to greet me. "Hello, Howard Frank." (He always addressed me as if I were a man, using my full name, to distinguish me from my father, Howard Hudson, and my grandfather, Howard Leroy.) "Let's walk up the mountain."

From the mountaintop where I'd spent so many happy evenings listening to those baseball games, we looked down at what was left of the town. Not a soul was in sight on the long, steep street. My uncle shook his head, then just stood silently beside me for what seemed like a long time. "Howard Frank," he said finally, "Chichester was a great town to grow up in, to live and teach in. Now it's a dead town. If Oat Morse were still alive, his bees would swarm on his chimney today."

"What can we do about it?" I said.

He thought for a minute. Then he said, "You like to write stories. Write a story about Chichester someday. Sometimes," he said after pausing again, "sometimes, telling the story of a place is all you can do to preserve it."

Now, as I sped on into the fall night, toward the big woods of Aroostook County, I found myself wondering who would write about the dying little fishing village of Lubec, Maine. Someone should, while there were still a few left who could remember when it was a community.

Notes from the Maine–New Brunswick Border

The sporadic Sunday traffic . . . suggested the casual intercourse between the two countries, here where the border is merely a turnstile on an open road and a time change; Maine is on eastern time, and New Brunswick is on Atlantic time.

— Marion Botsford Fraser, *Walking the Line: Travels along the Canadian/American Border*

9:00 A.M. The Georgia Pacific Paper Mill in Woodland, Maine. "From here north, you'll discover that the forest has shaped the lives of people on both sides of the border the way the ocean does down in Lubec," GP public relations director and former Maine timber cruiser Donna Peare tells me, and an hour later, I see exactly what she means. Since leaving her office, I've driven through a dozen densely wooded, uninhabited, unnamed Maine townships designated on local maps only by the capital letter "T" for "township" and a number: T29, T33, T46, and so on. Then farther back into northern New England's taiga-like forest, stretching one hundred and fifty miles north to Quebec and all the way west through New Hampshire and Vermont to New York State, broken by only a few mill towns and countless lakes and rivers. Big country!

NOON. Just across the border in McAdam, New Brunswick. Cameron Nason, a retired Canadian Pacific railroad man, has the disappearing-railroad blues: "See that tin locomotive on top of my weathervane? I made it myself. Soon enough, she's going to be the only locomotive left in these parts."

2:00 P.M. Lakeland Industries Center for Independence. There's no lake here, just the interminable Maine–New Brunswick woods and a long metal shed, where Roy Carvell operates a woodworking business with twenty handicapped employees from both sides of the border. "Not participants," he tells me at the top of his lungs over the shrieking saws, lathes, and planers. "*Employees.* I started the Center eleven years ago because of a terrible need in the area for jobs for handicapped adults. We began making picnic tables, switched to hardwood survey stakes and storage sheds, and now we're turning out decorative picket fences. I went gray in the first year, but our flexibility to be diverse has kept our heads above water. Though you might not guess it just driving through, there's as high a percentage of mentally disabled people up here in the woods along the border as anywhere else." Roy pauses and stubs out his third cigarette in ten minutes. This time he doesn't relight. "I have a disabled son of my own," he says almost offhandedly. "He works here himself now, and he's doing very well."

2:30 P.M. Vanceboro, Maine. Here in 1918 a German saboteur named Werner Horn attempted to blow up the Canadian Pacific trestle over the St. Croix River, which forms the Maine–New Brunswick border in this area. He succeeded only in shattering all the windows in Vanceboro and freezing his fingers, which he was trying to thaw out in the washroom of the local hotel when he was apprehended.

3:30 P.M. Orient, Maine, consisting of an official border crossing and a general store in the middle of the woods. The proprietor flatly informs me that in northern Maine the border is a nuisance, nothing more. I laugh and tell him that he speaks his mind. "You'll find that most people up this way do," he says. "The further north you go in Maine, the more people you'll meet who speak their mind. Wait until you get to the Madawaska Republic. Then you'll find out what independent-mindedness really is. They don't recognize *any* government up there, the U.S. *or* Canada."

10:00 P.M. Bedtime reading from the *Houlton Pioneer Times* in Houlton, Maine: "Melissa recalled for the jury how after [the murdered man's] remains were put in the woodshed it was still used for wood storage, and how her mother would sometimes tell Melissa to go to the woodshed to bring back wood for the stove. She said she would cry and her mother would laugh . . ." Enough! Tomorrow I'll see the Madawaska Republic, sometimes called the Louisiana of the North — but tonight it's a long time before I get little Melissa and her frightful trips to the woodshed out of my head and fall asleep.

The Louisiana of the North

Thinking they were in Canada, and not the area which is now Maine, the McKinnon brothers settled at the head of the Mattagash River. Oral history says they stopped where they did because one of the women had to pee. In truth, it may have been inaccurate maps. But as a result, even at its inception, Mattagash was a mistake.

— Cathie Pelletier, *The Funeral Makers*

In 1755 an event still regarded by some Maine Acadians as the Pearl Harbor of the eighteenth century took place in the tiny fishing and farming community of Grand Pré, Nova Scotia. The trouble began when Nova Scotia passed from French to English hands, and the local Acadians — French-speaking settlers who had come to the Maritimes over the course of the previous century — were herded at gunpoint onto deportation ships waiting to scatter them to the four winds. Some were sent back to France. Others were packed off to the inhospitable mountains of Quebec's Gaspé Peninsula. Several groups were marooned at random points up and down America's eastern seaboard, and still others were exiled to Louisiana, where they became known as Cajuns, an alteration of "Acadian." Two Grand Pré lovers, separated by force only to meet by chance years later, were celebrated in one of the most popular of all nineteenth-century American poems, Henry Wadsworth Longfellow's *Evangeline.* The little-known story of Maine's Madawaska Republic Acadians is equally incredible.

In the summer of 1766, a caravan of two hundred exiled Acadian families began walking north from Massachusetts toward Canada. Some of them settled along the Bay of Fundy. Others forged their way up the St. John River to Frederickton, today the capital of New Brunswick. But with the arrival there of hundreds of Loyalists fleeing the Revolutionary War, the diminished troop of Acadians was harried into the wilderness yet again.

In 1785 thirteen of the original two hundred families pressed on up the St. John to a beautiful section of the river high above Grand Falls, where the dreaded British deportation ships couldn't follow. Here they were welcomed by a friendly group of Micmac Indians, who called the lovely valley "Madawaska," Land of the Porcupine, and helped the fleeing Acadians settle in. The family names were Dufour, Cyr, Daigle, Hebert — the same names you will see in the Madawaska–Edmunston phone book today — and over the next two centuries they would populate the entire St. John Valley on both the American and the Canadian side of the river.

I met my first Acadian in Caribou, Maine, the southern gateway to the Madawaska Republic, early on the third morning of my trip. Fred Jackson had been recommended to me as one of northern Maine's smartest and most persistent game wardens, and a man whose own Acadian ancestry gave him a special insight into the population of the region.

In his early forties, in excellent condition, clean-shaven, with a brush cut and a restless, probing intellect, Fred Jackson looked like a high-ranking career Marine officer. To catch the poachers who ravage the big game of the Madawaska Republic, he has immersed himself up to his nose in the icy water of beaver runs, huddled under a thin plastic sheet in freezing November sleet, and clung to snowy granite cliffs for hours on end. He's tracked a mobster hit man from Quebec City through the night forest miles from anywhere, and tracked any number of sick and dying moose through miles of spruce bog. Warden Jackson has cornered armed desperadoes, combed wooden mountains for lost hikers, and talked about hunting safety with hundreds of Maine kids. And every day of the year he puts his personal safety on the line in defense of the woods and animals that he is sworn to protect.

"Most people who violate the game laws aren't very systematic," he told me over coffee as we sat at his dining room table looking out into the woods surrounding his house. "The great majority are simply opportunists who just happen upon an animal and decide on the spur of the moment to shoot it. That makes them hard to catch, but just last night I got lucky. Quite late, well after dark, I got a phone tip that some suspicious activity was going on in the back field of an abandoned farm near here. I drove over and left the truck out of sight up the road. The weather was lousy, but I could see what looked like a small light flickering through the rain near the woods at the far end of the field. The Department had just issued me a pair of infrared night goggles developed by the military, and with them I could plainly see three guys huddled around the carcass of a big moose. Right beside them was a backhoe they were going to drag him out of the woods with."

"Were you scared?"

"My heart was beating, fast, but no, I wasn't scared. If you think about being scared, you can't do this kind of work. You have to focus on your job. I crawled up through the field on my belly, being as quiet as I could. When I was about ten feet away from the poachers, I stood up and flashed on my light and said, 'What's happening, boys?' They were dressing out the moose with their hunting knives. I'd literally caught 'em red-handed!"

Fred Jackson told me that when he first started out as a warden, several old-time Acadian poachers from over the border in Canada played cat-and-mouse games with him, as they had with his predecessors. "These were families that had sneaked over the line to poach moose and deer in the U.S. for generations," Fred said. "Their ancestors were here long before the border, so they didn't recognize it at all, except as a minor inconvenience. Those old-time Acadians really believed that God put the moose and deer and bear here for their use. I can understand that. After all, I'm half Acadian myself. But that's just one side of the coin."

"What's the other?"

"I'm half Acadian," Fred Jackson said with a big grin. "But I'm all game warden. Enjoy your visit to the Madawaska Republic."

The moment I laid eyes on Ernest Chassé of Madawaska, I knew what Warden Fred Jackson had meant by an old-time Acadian. He wore a vivid red-and-black plaid shirt and fire-engine-red suspenders, had sharp dark eyes, and a shrewd and jovial countenance, and greeted me in a rumbling baritone voice I could listen to all day long. Stick a short pipe in Ernest Chassé's mouth and he'd be a dead ringer for the life-size woodchoppers carved from pine that you see outside the shops of Acadian *artisants* throughout New Brunswick and Quebec. At eighty-three, he had the vitality of a man half his age.

"Come in, sir, come in," Ernest said. "Welcome to the 'Louisiana of the North,' as I like to call the Republic. If Geraldine, my wife, were here, she could tell you all the history of the first thirteen Acadian families. I myself have only one small story. Seventy-one years ago, when I was twelve years of age, the local priest confirmed me into the church on the spot by the St. John River where the monument to the thirteen founding families now stands. For generations, all the Acadians of the Republic had been confirmed there. Very well. Much later I bought that land. I wanted to grow potatoes on it. But I was told that was impossible. People assured me that nothing would ever grow on the site where the original thirteen had landed. In fact, there was one piece of land by the river, perhaps an acre, where it is very true that all the potatoes were small and scabby. So I said to myself, 'Well, Ernest! We will just see about this matter.' "

Ernest Chassé laughed and nodded, relishing what came next. "You must understand," he continued, "that I am not a superstitious man. But I am very determined. I wanted to show that good potatoes could be grown there. You see, I remembered from when I was a boy that an old building, an Acadian barn, had stood on that place. That was the true reason the potatoes would not thrive. Potatoes do not like to grow where manure has been. So I called in the county agent and he helped me treat the soil. The first year the potatoes were still small and scabby. We treated the soil again, sweetening it with quantities of lime. The second year, better. By the third year I was harvesting big, sound, smooth Aroostook County potatoes. I had proven it could be done and that is all I wanted to do. The following year, with Geraldine's approval, I gave the land to the local historical society to make a monument; and now, sir, while there is still good light to see by, you must make a visit to that monument, where you will read on white marble the names of the original thirteen Acadian settlers of the Republic, just a stone's throw away from the river that was once the site of the biggest log drive in the world."

It was the most dangerous work in the North Country, bar none. But each spring in late April or early May, when the ice went out of the upper St. John River and its tributaries, Madawaska Acadians from backwoods farms and tiny nameless hamlets on both sides of the river came flocking by the hundreds to do it. They ranged from fifteen or sixteen to sixty and older, and they were all eager to risk their lives daily for the next two or three months to work as rivermen on the annual spring log drive down the St. John.

In the peak years of the St. John drive, around the turn of this century, rivermen took one hundred and twenty-five million board feet of lumber a year down the four-hundred-mile-long river, from its source deep in Maine's Aroostook County to the tidal estuary of New Brunswick's Bay of Fundy. The drive sometimes stretched out over seventy-five miles of river and required fifteen hundred men to manage. At home on the farm or in the little mills where they worked the other nine months of the year, these Madawaskans might not have looked like extraordinary men. But up on the drive, they were something to see. They wore layers of bright wool shirts, and on their heads tuques brighter still, or slouch hats with rakish floppy brims to keep off the rain and sleet. Their tight pants were tucked into the tops of high, calked boots, each boot bristling with sixty-three razor-sharp steel spikes three-quarters of an inch long. They worked from first light to full darkness, often up to their waists in rushing, ice-filled water, and they slept on the bank in their wet clothes wherever they happened to be when night fell.

Madawaska rivermen kept their paper money buttoned in their shirt pocket, but it rarely stayed there long. When they hit the Republic towns of Allagash and Fort Kent and Madawaska, they spent what they'd earned with the prodigality of sailors on shore leave, playing as hard as they worked and fighting as hard as they played. With bare fists and teeth and those formidably spiked boots, they fought up and down the muddy main streets of towns all along the St. John, into and out of the taverns and whorehouses, sometimes with the men on their own crew, sometimes with men from another outfit, and sometimes with loggers and rivermen from the opposite side of the river. And in 1839 their brawling very nearly precipitated a full-scale war between the United States and England.

The conflict involved the tremendous stands of virgin timber in the upper St. John Valley and the rightful ownership of the disputed land it grew on. Both Maine and New Brunswick had laid claim to the territory; and lumbermen on both sides of the river were cutting trees to beat the band and fighting with each other all the way downstream to the mills. Fortunately, a level-headed American general named Winfield Scott marched up from Augusta with a contingent of well-disciplined Maine militiamen and worked out a temporary truce with the irate Canadians. In 1842 a permanent treaty settling the dispute was negotiated by Senator Daniel Webster and the popular English diplomat Lord Ashburton, dividing the twelve thousand square miles of disputed territory approximately in half, with Maine getting about seven thousand to Canada's five thousand.

Still, driving up along the St. John toward Fort Kent that evening, I couldn't seem to shake the impression that Maine's wild and beautiful Louisiana of the North, which juts well to the north of Quebec City, rightfully belonged in French Canada. "The Madawaska Republic is really just one big community on both sides of the border," a Fort Kent businessman told me that night over dinner. "My father, a Canadian, married an American. My wife's an American, but she works across the river in Canada. My brother's an American, but his wife's a Canadian — you get the idea. For years the customs people have just waved us back and forth across the border, which may be unique in the world today in that for us residents of the Republic, it simply doesn't exist, except on paper. The fact is that nobody up here's taken it seriously since the Webster-Ashburton Treaty, a hundred and fifty years ago. On the other hand, we have to wonder how seriously the *rest* of the state and country takes us. Down at the state capital, in Augusta? They're aware that the Madawaska Republic exists, all right. But to tell you the truth, they seem to feel that their main responsibility to us is to save us from ourselves, and that's infuriating."

He paused. Then, smiling but serious, he said, "We'll render unto Washington and Ottawa and Augusta and Frederickton what's rightfully theirs, but not one speck of tribute more. As for the border, I don't see any border, do you? Just a beautiful country with a river running through it."

Flying the Border with Ti René

Flying a bush plane is almost as good as sex.

— Ti René, Québecois bush pilot

Buoyed up on its bright aluminum pontoons, Ti René's ancient yellow de Havilland Beaver bush planed taxied in wide ellipses in front of his camp on the remote Quebec-border pond he's used for the past forty years as a base. Then we were ripping over the water straight into the early morning wind and lifting off the surface like some primitive reptile trying to fly for the first time. The vast woods and water of northern Maine and southern Quebec spread out below us like a gigantic blue and green jigsaw puzzle, stretching as far as I could see in every direction.

I'd met Ti René late the night before in a Quebec roadhouse with a wide open topless bar attached, where I'd stopped for a beer. The bartender had recommended him to me as an "outlaw" Québecois bush pilot, a notorious smuggler, who might take me up for a bird's-eye view of the border the next morning.

"No problem," Ti René said between joking conversations in French with strippers young enough to be his daughters. "Just don't write where you found me, or Ti René's last name."

René gave a great booming laugh. At six foot six and somewhat over three hundred pounds, he seemed to derive great pleasure from the irony of his nickname, *Ti,* which is French Canadian for "Junior" or "little" — Little René!

Now, as we continued to climb higher, I pointed down at a narrow cut through the trees far below. "Is that the border?" I called out over the engine noise.

"The what?" René shouted.

"The border."

"Oh, that —" with a contemptuous wave of one huge hand. "Yes. But the air is free, there is no border up here. Still, I will give you a good look at the line on the ground, eh?"

With a wicked grin, René abruptly rolled the de Havilland over onto its side, pointed his left wing tip at the boundary and plummeted down toward it. The landscape tilted sickeningly as it rose to meet us. But René laughed and whacked my knee with his ham of a hand and shouted, "Look off at the horizon and the dizziness will pass! Ti likes to play with the plane, eh?"

On the way back up, after tracking the border a hundred or so feet above the trees for a few miles, then losing sight of it altogether in a morass of spruce swamps and beaver ponds, Rene said dismissively, "Canada, the U.S., the U.S., Canada — it is all about the same, right? You know what the line here is called: the friendliest border in the world. I think so myself."

The border, as it turned out, has run through René's family like a genealogical line for the past three generations. Like many another turn-of-the-century Québecois farmer, his grandfather emigrated to northern New England and bought a small, rundown dairy operation. He milked a dozen cows, did a little carpentry and lumbering for his neighbors, played his fiddle at local barn dances, and, during Prohibition, augmented his income by running Canadian whiskey across the line in a Model A with a homemade truck bed attached. Then grandpère had an inspiration. He traded one of his woodlots for a broken-down biplane, coaxed it into running order, removed the wheels and added a pair of pontoons, and began smuggling whiskey over the border in considerable quantities.

Was coming to the States a good move for his grandfather? Oh, sure, René said. Back then a hard-working French Canadian immigrant family could, with a bit of ingenuity, flourish in the U.S. And grandpère had had more than a tad of ingenuity. Both René and his father were educated with money his grandfather saved from those whiskey runs in the biplane. Now, though, he told me, Québecois farmers were staying put. With their handsome government subsidies, dairymen in Canada were doing much better, for the most part, than their beleaguered counterparts in New England.

René's grandfather began taking him up in the plane when

René was five years old. Since then he has flown thousands of wilderness sorties on both sides of the border that he pretends to ignore. He's ferried into the bush game wardens, fire wardens, timber cruisers, guides, newspaper reporters, fly fishermen, backpackers, emergency medical teams, firefighters, paper company executives preparing publicity brochures, and "Earth First!" activists eager to photograph the devastation wreaked by the paper companies' clear-cutting. He's flown fish-stocking planes and big cumbersome "water bombs" — bright orange firefighting planes that lug thousands of gallons of water in gigantic bulbous compartments attached to their fuselages. He's put in several stints as a crop duster over Aroostook County potato fields, has flown nonstop, except to gas up, from dawn to dusk, day after day, searching for lost hikers, and has hired out his services to U.S. Border Patrol agents to comb the border terrain for illegal aliens and to Mounties in search of escaped murderers.

In the winter, Ti René swaps his pontoons for a pair of wide skis and lands and takes off from frozen lakes and ponds, supplying outback trappers, timber cruisers, and lumberers with necessities. And though some customs authorities raise an eyebrow when you mention René, who has been known to carry a few undeclared cases of Molson's Export or Seagram's over the line, he has an unimpeachable reputation as a skillful and safe seat-of-the-pants bush pilot, who can fly anywhere without instrumentation, navigating entirely by maintaining sight with ground features and landing wherever a bush plane can land, as well as a few places where, according to the laws of physics, one shouldn't be able to. During his forty-five-year flying career, René told me, he's landed on some ponds so small that, in order to get the necessary taxiing distance to take off again, he's had to hitch his plane to a shoreline spruce tree by a stout hawser, rev up the engine until the hawser hums with the tension, then pull the slipknot and use the plane's lurching leap forward as momentum to take off from water "too small to take off from" — a strategy he learned from his grandfather.

* * *

This morning I accompanied René to Portland, where he picked up a rebuilt log-skidder transmission (and six cases of American beer) to deliver to a lumber camp in Quebec.

"Does anyone ever bother you about bringing booze over the border?" I shouted once we were aloft again.

René laughed. "No problem. Look, Ti René is no criminal. A judge would laugh such a case out of court, you know, arresting René over a few bottles of beer, a small present for his friends across the friendliest border in the world."

As we traced the border back toward René's base, I asked him if he thought Quebec would sooner or later secede from Canada. He shrugged. "Who knows? Very possibly. But only consider, my friend. In many ways Quebec has already, for all practical purposes, seceded. French is now the official language of all our public schools and the language spoken in the Parliament buildings in Quebec. Furthermore, as you drive through most of the province, you will note that the flag you see flown, including over government buildings, is the fleur-de-lis, the Quebec flag. So, now that we have such great cultural and linguistic and yes, political, autonomy, it may not be wise to cut our last economic ties with the rest of Canada. True, the federal taxes are steep. But we have wonderful medical care, care for the elderly, educational benefits — both my brother, a doctor, and I attended the University of Quebec for next to nothing, and our grandfather paid our living expenses while we were at college. Our farms are alive and healthy, our social programs among the best in the world. Driving through Quebec you will not see the sad poverty that characterizes the little mill towns just across the border in northern New England. In Quebec, the yards have colorful flowers in them. In Maine and New Hampshire and Vermont, many have junk cars on wood blocks and old refrigerators! All this is something we want to jeopardize for national pride, to call ourselves a nation? René hopes not! Will we, though? He fears that sooner or later we may. If so, it will be a mistake."

Shortly before we landed, I asked René what he liked best about flying a bush plane for a living. "Why do I do this?" he said with a chuckle. "Because here, you see, I am free. Look. The law says you cannot fly a bottle of beer over a line that scarcely exists except on a map. Now what does such a law mean to me? Up here in the air" — René snapped his fingers — "not that much. Man's law! Petty regulations. Now God's law. That René makes every effort to obey to the letter. He is not always successful. But that is what he strives for."

I asked René for an example of God's law. He laughed and said he knew of just one: "Treat other people the way you'd like to be treated."

On that note, without warning, René lifted the big metal steering rudder up in its housing and flipped it smack into my lap. Before I knew it, I was at the controls, feeling all the throbbing power of that ancient de Havilland Beaver in my hands.

"Almost as good as sex!" René hollered. "But not quite."

After a few minutes he took the controls back. "So," he laughed, "you have flown a bush plane, and gone on a smuggling run with Ti René, and lived to tell about it. To write about it even. Well, write this: This wilderness we have flown over? This border country? Write that it is a hard place to make a living, but a good place to live! And write that it is beautiful, eh? Very beautiful, but ever so fragile as well. Write that once gone, it does not come back again."

An Allagash Guide

We were now fairly on the Allegash River, which name our Indian said meant hemlock bark.

— Henry David Thoreau, *The Maine Woods*

The following afternoon, in the tiny village of Jackman, just east of Moosehead Lake, I had a piece of traveler's luck. At the general store where I stopped for a Coke, I happened across a recent profile from the local paper of a Mr. Harry Hughey, a long-time Allagash River guide and Jackman resident, who at ninety years old still took fishing parties into the Maine woods. I had driven here on Maine's wilderness Golden Road through some of the same Allagash territory my father and uncle and I had fished thirty years ago. On the surface much of the area seemed about the same, though in other places whole mountainsides had been scalped of their timber by clear-cutting — and I couldn't think of a better interpreter of the changes these big woods had undergone than a veteran Maine guide. The proprietor gave me directions to Mr. Hughey's house, on the bank of the Moose River at the end of a woods road just west of town, where I found the guide getting up his woodpile for the coming winter.

Entering the tenth decade of his life, Harry Hughey looked like an ex–welterweight boxer who's taken great pains to stay in shape since retiring from the ring. The son of Scottish immigrants from Canada, he began guiding as a teenager at his father's fishing and hunting camp on Boundary Pond, on the nearby border with Quebec. Since then he'd made the trip down the Allagash dozens of times; seen tens of thousands of big brook trout caught and scores of big deer shot; paddled and carefully documented the entire northern Maine route of his favorite author, Thoreau, whom he referred to almost reverently by all three of his names, Henry *David* Thoreau, with an emphasis on the middle name; run a trapline that kept him out in the winter wilderness for weeks at a time; and cruised and scaled timber all the way from the Madawaska Republic to Moosehead Lake.

It was late afternoon when I arrived, and getting chilly. Harry had a small wood fire going in his workshop, where we began our visit sitting in two straight-backed kitchen chairs near the potbelly stove. On the walls around us hung Harry's deer rifle, shotgun, fly rod, an assortment of axes, saws, and carpentry tools, a pack basket, and some black-and-white framed pictures of long-ago clients with deer and trout. "Those are native trout, also called brook trout or squaretails," Harry said. "Sports would come all the way up here from New York and Boston to catch 'em. They'd take the train to Skowhegan, then catch a stage to Greenville, at the south end of Moosehead Lake. My dad and I would meet them there in a buggy for the twelve-mile trip through the woods to our camp."

Talking a little shop, Harry said that in those days the most popular flies for brook trout were the gaudy old Silver Doctor, Parmachenie Belle, Dark and Light Montreal, and Red Ibis, often fished three at a time, with a lead fly and two dropper flies at two-foot intervals up the leader. When the trout were feeding well, it wasn't unusual for fishermen to catch a "double" or even a "triple" — three fish at once! The best fly rods were handmade Fenwicks, Thompsons, Paynes, and Hardeys, with Orvis coming along a bit later, in the 1920s. Harry was quick to acknowledge the superiority of the recently developed Muddler streamer as the best all-purpose trout fly ever tied and the advantages of today's sensitive and strong graphite fly rods over the old bamboo wands. Nor was he, strictly speaking, a purist fly fisherman: "When they won't rise to a Muddler, I carry my flies in a gallon pail," he told me. "I prefer a fly, but I've found that Maine brook trout taste just as good when they're caught on worms."

How about canoes, I asked. Did he have a favorite make? Harry told me that for the long Allagash trip, the twenty-foot canvas-covered White canoes were better than the renowned Maine Old Towns. While the Old Towns were broader of beam and would stand a heavier sea on the big lakes, the White was an easier paddling craft. It weighed ninety pounds and was what he called "a pretty good carry," meaning manageable to lift over his head and tote on foot over the portages, some of which were four miles long through incredibly rough terrain.

Like Henry David Thoreau, Harry jumped off from Greenville for his Allagash trips. He paddled up to the top of Moosehead Lake, made the short portage over to the West Branch of the Penobscot River, then crossed Chesuncook, Eagle, and Churchill lakes before spinning on down the hundred-mile-long Allagash to the St. John — leaving plenty of time for delicious side excursions to remote ponds and streams teeming with big brook trout. "Those were good trips," Harry said. "It was a more leisurely era back then, in the twenties and thirties. After we got out in the bush, there wasn't any bus to catch. Oh, once in a great while I'd get a crabber, somebody who'd spend the whole month grousing. The next year when he wrote to me about coming back, I'd be busy. But most of the sports I guided were wonderful people, people who loved the woods the way I did."

I asked Harry if he remembered the big log drives like the one over on the St. John. He said he ought to, he'd gone on some of them. In 1924, when Harry was nineteen years old, he'd worked on the last drive of long logs on the Moose River, "swamping, tending out, and hanging the booms." A swamper, Harry explained, walked the riverbanks as the drive moved down with the current, using a ten-foot-long pike or "pick pole" shod with an iron point to dislodge logs hung up in backwaters. Tending out was fun. To do that, Harry stood at the bottom of the rapids or just below a wooden driving dam and used his pick pole to guide the giant logs sluicing past him into the swiftest current. Best of all was hanging the booms — chaining together the huge necklace of boom logs that enclosed the thousands of saw logs and was towed by steamboat across the many lakes and ponds along the length of the drive. To string the booms together, Harry worked from a batteau, a thirty-two-foot-long wooden boat manned by four oarsmen, with a helmsman at the tiller. Filling the booms with logs coming into a pond from the inlet was "God's job," however. For that you had to wait for the wind to be in your favor.

Was river driving dangerous work?

"Of course it was," Harry said. "You had to know exactly what you were doing every minute; but that was true of guiding and timber cruising, too. Some of those big North Woods lakes we crossed in canoes have very heavy seas from time to time, waves four feet tall and taller. Henry David Thoreau's Indian guides were afraid of them, and with good reason. The trick was to stay off big water in a wind. We'd slip along near shore from cove to cove the way Henry David did, either that or cross the big pieces of open lake at night when the wind was more apt to be down. Quite often, too, we'd hit out onto a lake before first light and stop to cook our breakfast at dawn on the far side. I never did capsize a canoe, but lots of men drowned on those lakes and rivers — log drivers, guides, sports, trappers going through bad ice in the spring. As I say, you wanted to know what you were about.

"You didn't want to wander off into the woods alone, either. I had to watch my sports very closely, know where they were at all times. Back then you could walk thirty or forty miles through the Allagash wilderness and never strike a road. Not so much as an old log trace, even. Even experienced woodsmen could get into trouble. To this day up there, you still want to take a compass when you go to the toilet!"

I asked Mr. Hughey how he got his first guiding license. Was there a test? He smiled. "No. Back then the local game warden just gave you a license if he thought you were all right in the woods. No fuss or nonsense."

During the winters, Harry had divided his time between running his wilderness trapline and cruising timber. At one time or another, he'd tramped over most of northern Maine in search of the last stands of big timber, seeing the same country on snowshoes that he saw summers by canoe, when it looked altogether different, "two entirely separate worlds."

"Do you know what a Penobscot rule is?" Harry asked me.

I'd read about such a tool in Robert E. Pike's wonderful anecdotal history of logging and river driving in northern New England, *Tall Trees, Tough Men*, but I'd never seen one. Harry took me out to his woodshed, where he showed me a measuring stick with squared-off sides, about four feet long and covered with numbers. This was his Penobscot rule, which he'd used for scaling the number of board feet of lumber in a log.

He turned the old rule, reinforced around the middle with black electrical tape, in his strong hands. "I wouldn't mind having all the wood this has scaled," he said. "I expect it would make a pile as high as that mountain over across the river."

Then he gave me a rueful grin. "Of course, that's just the trouble with this country today, you know."

I looked at Harry Hughey inquiringly.

"Cutting off whole mountainsides," he said. "They say clear-cutting doesn't hurt the North Woods, but the fact is that it's the worst thing that could ever be done to them. Every time it rains, the runoff from the cut-over slopes turns the streams the color of muddy coffee, the fish die, whole *woods* die. It's a disgrace. In my lifetime I've seen northern Maine go from a near wilderness to a pale reflection of what it once was. On the other hand, *any* woods is better than no woods, and the paper companies that own most of this land do keep it in trees and make it accessible to the public. It isn't what it once was, but what is? I count myself lucky at ninety to be able to still get out in what's left of it."

Brook Trout Fishing

I fish mainly because I love the environs where trout are found: the woods; and further because I happen to dislike the environs where crowds of men are found: large cities; but if, heaven forbid, there were no trout and men were everywhere few, I would still doubtless prowl the woods and streams because it is there and only there that I really feel at home.

— Robert Traver, *Trout Madness*

Time and again, on the journey that lay ahead of me, in upper Michigan and Minnesota and Montana and Washington, and on the Canadian side of the border as well, fishing seemed to be the universal North Country sport linking the people I met with their territory, their history, and one another across the generations. It was as though their fishing lines, reaching down into the icy streams and lakes of the border country, connected the natives of these places to their geographical regions in a tangible way, as I had been connected to my own sectors of the North Country, and to my family and friends as well, by the fishing lines looping and arcing back into my own past. Because fishing, particularly for Mr. Harry Hughey's beloved brook trout, has been a passionate lifelong avocation for me as well.

I knew, too, why Harry Hughey kept his lovely old bamboo fly rod displayed on wooden pegs in his workshop instead of packed away inside an aluminum case. Partly, of course, he wanted his rod near to hand because he still used it nearly every day of the trout season. But that fly rod also had a transcendent talismanic and autobiographical significance, as the bamboo fly rod I was carrying in the trunk of my car did for me.

Two of my sharpest boyhood memories involve fishing. In one I am standing on a chair at the kitchen sink next to my father, who has just come home from fishing the Catskills mountain stream behind our house. "Ready?" he says. And he lifts his wicker fishing basket and spills out into the sink a limit of shimmering brook trout, still packed in the damp, woodsy-scented ferns he's lined his basket with. He notices the expression on my face, in my eyes, and he smiles and says, "Next year, buddy. Next year I'll show you how to catch these."

So it's the following June, I've just turned five, and I'm standing beside my father on a granite culvert over that same brook. Though my father is an inveterate fly fisherman, he knows exactly how to teach a small boy to fish for trout and knows that doing so has nothing at all to do with flies. He's rigged his Orvis bamboo rod with a stout leader, a number-ten Eagle Claw hook, a split-shot sinker, and a lively red garden worm, which he flips into the spillway below the culvert. He casually hands the pole to me and, less casually, I take it in both hands. Almost immediately the tip jerks down sharply — a brand-new feeling, a good one! And at that instant, I know that I am connected to something down there. Something important.

Dad laughs. "Go ahead," he tells me. "Yank him out."

With both hands I give a great heave and derrick out of the small culvert pool my first brook trout, sparkling red and gold, with forest-green patterns on his back and delicate azure halos rimming the crimson speckles on his sides. The trout is well hooked — he'd gone flying right over my head! — and so, forever more, am I. And although I have no doubt caught tens of thousands of brook trout in hundreds of streams and rivers since that day, the surprise of that first trout's sudden *tap-tap-tapping* strike, his gorgeous colors, the sunny mountain morning, and the companionship and approbation of my dad, the best trout fisherman I would ever know, are still clear and fresh in my memory today.

Upstream from that culvert, where the brook ran through a series of brushy meadows and mixed hardwoods and evergreens, you could hop across it in one jump in the summertime. Here, sometimes alone, sometimes with my friend Lennie from the village, and sometimes with my father and uncle, I honed my fishing skills. Higher up still, in a notch between two steep mountains known as Ox Clove, was a series of wide, smooth, gently sloping, maroon-colored ledges, over which the stream ran scant inches deep. These ledges were known locally as Red Rocks. Below each one was a small pool, and each pool usually contained several good brook trout. At Red Rocks, under my father's and uncle's tutelage, I caught my first fish on flies. Like Dad and Uncle Reg and Harry Hughey, I used three flies, a lead fly and two droppers. "You pick them," my uncle said, opening his fly book to an exciting array of big, old-fashioned, colorful wet flies. "Anything bright. Brookies love anything bright."

They did, too. There was little technique to catching those eight- and nine-inch speckled trout at Red Rocks on wet flies: toss some bright-colored flies into the pools below the ledges, let them drift down the current, and bingo! I had a fighting fish on my line before I knew it. If I tangled my flies on a backcast or horsed a fish in too fast and lost him, Dad and Uncle Reg never criticized me. If, on the other hand, I caught a bigger fish than usual, the most they'd say by way of compliment was "good fish." For me that was enough. "Good fish" meant "good job" and more. It meant that I was becoming a brook trout fisherman, like Dad and Uncle Reg.

When I turned twelve, I began accompanying my father and uncle on their annual trout fishing expeditions to northern Maine and to the Laurentian Mountains of Canada. Until I began exploring the northern reaches of Labrador and the Ungava Peninsula of Nouveau-Québec with my own son, thirty years later, Laurentides Park, north of Quebec City, was the wildest country I'd seen, and the best. Here for a week to ten days each August, we'd canoe authentic wilderness rivers and lakes, often encountering not one other person on our entire trip. The trout in those frigid Canadian waters were both large and abundant; yet neither my father nor my uncle made any kind of mystique out of catching them or teaching me to catch them. Still, there was a code, and though it was mostly unspoken, I absorbed it early on. You always put all the trout back in the water alive except for a few to eat. You didn't count your trout or call attention to their size or weight. You took time to watch and enjoy seeing your partners catch trout and to enjoy the natural setting where the fish lived, and the otters, moose, bears, wolves, eagles, and ospreys that lived there with them.

At the same time you could, if you wanted, use brook trout fishing as a standard of comparison, as my father and uncle did when they agreed that catching a big squaretail in a remote Canadian lake was even better than hitting the game-winning home run for the town team in the county playoffs. They agreed as well that except for a select few men they didn't even know — Dashiell Hammett, Raymond Chandler, Ted Williams, and Ernest Hemingway come to mind — they would never take anyone else but me brook trout fishing with them.

Here too, in those distant Canadian mountains visited only by a few fishermen, I felt totally at home, felt a sense of belonging that I would subsequently experience only in remote rural or wilderness settings and only with family and the closest friends. I had no doubt at all that the North Country was where I wanted to live.

When I graduated from college, my father gave me his Orvis bamboo fly rod and bought himself another one. I wasn't ready to break out that rod on this trip yet. I'd wait a while to do that, and the anticipation would be good too. In the meantime, however, I knew that Harry Hughey was a man my father and uncle would have taken brook trout fishing, and I knew that meeting this Allagash guide, even in a wilderness that was just a pale reflection of its former self, had made my trip worthwhile already.

Notes from Indian Stream Territory

The people inhabiting the territory called Indian Stream Territory do hereby solemnly and mutually agree with each other to form themselves into a body politic by the name of *Indian Stream* and in that capacity to exercise all the powers of a free, sovereign, and independent state so far as it relates to our internal Government.

— Indian Stream Territory Constitution,
Indian Stream Republic, 1832–1842

8:00 A.M. The Eastern Townships of Quebec. Heading west toward northern New Hampshire, I have a day's drive in which to reflect on my experiences in Maine. Like my home in the Northeast Kingdom, much of Maine's remote border country still has a frontier feel to it today. But besides the colorful histories of its far-flung North Woods enclaves and their inhabitants, I'd discovered an interesting duality in the way the people view the vast forests that traditionally have given them their living. Men and women like the hardscrabble Aroostook County potato farmer Ernest Chassé, Allagash guide Harry Hughey, timber cruiser turned paper company executive Donna Peare, ex–railroad man Cameron Nason, sardine fisherman Elisha, "outlaw" bush pilot Ti René, and game warden Fred Jackson all love the rugged locales where they live and with which they're so closely in touch; but they never underestimate the natural and economic forces that constantly jeopardize their livelihoods there. Inherent in their endless battles with the weather, the sea, the rivers, the forests, and outside regulations and markets and bureaucratic decisions is an ongoing dramatic conflict that seems to have characterized life in the North Country for the past two centuries. What's more, it seems to me that most of the North Country natives I've met enjoy, even relish, this conflict.

* * *

EVENING OF THE SAME DAY. The Canadian customs station on top of the ridge marking the border between Quebec and New Hampshire. To the south, the evergreen forest of the Indian Stream Territory of northern New Hampshire stretches away and away, concealing First, Second, and Third Connecticut lakes, where a few miles south of the border the Connecticut River rises to flow nearly four hundred miles before emptying into Long Island Sound. To the north, starting just a quarter of a mile away at the foot of the ridge and continuing as far as the eye can see, is the pastoral countryside of Quebec — white stone and stucco farmhouses with bright orange roofs, extensive flower gardens, neat rectangular hayfields and cornfields, and blue-clad plaster Madonnas gazing placidly out over the landscape from upended bathtubs. Ti René was right. Geographically, culturally, economically, the differences between the two countries here on the New Hampshire–Quebec line are astounding. I don't know this yet, but in no place along the entire border, from Maine to the Pacific, will the contrast between the Canadian and U.S. sides of the line be so abrupt and noticeable.

7:30 P.M. On through the dusk toward Pittsburg, New Hampshire, where the upper Connecticut resembles a big western trout river, and the woods press tight to both sides of the road for twenty miles. In this forbidding region in 1832, several hundred Yankee, French Canadian, and Indian loggers, river drivers, homesteaders, trappers, and hunters banded together to declare the total independence of northern New Hampshire from both the United States and Canada, raising their own militia and writing their own constitution. Known as the Indian Stream Rebellion, this impromptu border-country insurrection lasted until 1842 when, as part of the Webster-Ashburton Treaty, the disputed tip of New Hampshire passed peaceably to the United States. "Independent" is the key word in the Indian Stream Constitution, I think: "a free sovereign and *independent* state." As my Canadian-born editor friend Chris

Braithwaite said to me before my journey, independence, rooted in local land ownership and local government, seems to have remained the chief objective of northern New Englanders to this day, the common denominator linking all of the people I'd met thus far with one another, with my friends and neighbors in Vermont's Northeast Kingdom and, for that matter, with me.

A Green Mountain Spy Story

It's no secret from anybody that both borders of this U.S. are sprinkled with sections that are as lawless now as they ever were in the old days.

— Dashiell Hammett, *The Continental Op*

"You won't find a more beautiful town on the entire border than Derby Line, Vermont," said Larry Curtis, a former criminal investigator for the U.S. Customs Service, early the next morning as we sat in the living room of his house just off the main street of that handsome village. Larry, a plainspoken man in his fifties who looks right straight at you when he talks, was born and raised in Derby Line and spent nearly all his career as a customs investigator working here. But the story he was about to tell me really began when he was just out of college and teaching "all subjects" in a tiny nearby high school, when the board of education asked him to coach the girls' basketball team — for no additional pay.

"I'd probably have done it for a couple of hundred dollars extra," Larry told me. "I was only making thirty-six hundred dollars a year, and two hundred more would have come in handy back then. But the school board wouldn't offer me an extra penny, so I said no. Then two things happened. In order to find someone to take on the basketball job, the board had to import a coach from outside the town and pay him eight hundred dollars. At about the same time they hired a new janitor for five thousand a year. This guy had just gotten out of the Vermont State Prison, and to me, that was the last straw. They were willing to pay an ex-con fourteen hundred dollars more a year for sweeping the classrooms than I was earning to teach their kids! I'd been working summers for the Customs Service, and you can bet that when a full-time inspector's job opened up shortly afterward, for considerably more money than I was making teaching, it didn't take me long to latch on to it."

After five years as a regular inspector working the border, Larry Curtis was promoted to the position of criminal investigator. An adventuresome man with a taste for action, he liked his work and became very good at it. But neither he nor anyone else could have dreamed that it would eventually lead him out of the most beautiful town on the entire border on an odyssey to a dozen different countries and into a labyrinth of international gunrunning, murder, and political intrigue as exciting and complex as the plot of any spy novel.

One cold spring day in 1978, Larry's supervisor in Boston called him with a bizarre story. A large container being loaded onto a ship bound for South Africa from the tiny Caribbean nation of Antigua had fallen off a crane, burst open, and scattered what was apparently a covert shipment of gigantic howitzer shells galley-west all over the ship's hold. The shells were quickly traced to a U.S. firm called Space Research, Inc., with headquarters in (of all places) northern Vermont. Space Research was supposedly studying weather patterns, but the spilled howitzer shells told a very different story. Citing the United States' and United Nations' embargoes on arms shipments to South Africa, the Antiguan government had filed a formal complaint with the U.S. ambassador, who had requested an immediate customs investigation.

"My boss called me on Friday, and wanted me in Barbados the following Monday morning," Larry told me. "Air connections from the Northeast Kingdom were so bad that it took me until Wednesday to arrive, but from the moment I stepped off the plane I worked on nothing else but the Space Research case for the next two years."

In Antigua, Larry was met with what he described to me as "an incredible runaround" from both Space Research employees and bribed Antiguan officials. Encountering resistance at every turn, he made his way to Space Research's oceanside testing site, pad and pencil in hand, and began systematically inventorying the stockpile of howitzers, shells, and radar tracking equipment that had been shipped into the country through Antiguan customs. When he came up with huge unaccountable shortfalls, the arms company claimed that they'd used up thousands of shells in testing — then dumped their worn-out howitzers in the sea.

Back in Vermont, Larry immediately headed for the Space Research headquarters, just north of the popular ski resort at Jay Peak. Here, on the Vermont-Quebec line, he discovered a munitions compound where, he was certain, an affable-appearing Canadian-born scientist and ballistics expert named Gerald Bull was manufacturing and smuggling out of the country a sophisticated and powerful long-distance artillery system capable of hitting targets more than twenty miles away with pinpoint accuracy and frightful results.

"From the day I got back from Antigua I was out there at the testing site, pestering Bull night and day," Larry said. "Eventually I located documents tracing the destinations of every shell and weapon that went into that compound. The way it worked, Bull would bring empty shell casings he'd purchased at U.S.-government-operated companies onto the compound from the American side, load them with warheads, then truck them out the Quebec side without passing through customs, to be shipped out of Canadian ports like Montreal. It was a geographical arrangement that never should have been allowed, but the CIA and other U.S. agencies that wanted to give under-the-table assistance to South Africa in its war with Cuban-backed Angolan rebels were greasing the skids for Bull. It was a CIA agent who put South Africa in touch with him in the first place."

Who was this mysterious scientist, Gerald Bull, who had masterminded both the advanced new weapons system and the ingenious method used to smuggle it out of the country? Born in Canada in 1928, Bull was the youngest of ten children of an English-speaking father and a French mother. A mathematical prodigy, he received his doctorate in aeronautical engineering from the University of Toronto at the age of twenty-two and went on to do postdoctoral research at McGill, in Montreal. There, in a failed experiment, he literally blew up a university laboratory. At McGill, however, Bull also managed to get his hands on the original blueprints for Germany's "Paris Gun," the gigantic World War I cannon that had bombarded Paris from seventy-five miles away.

Some years later, using the Paris Gun as a model, Bull welded together two gigantic naval cannons from a mothballed battleship to construct his own huge "Supergun" in the secret compound deep in the Vermont mountains. He called his creation Long Tom — a one-hundred-foot monstrosity originally designed to hurl weather satellites into orbit, but in fact capable of launching rocket-assisted shells the size of Volkswagens, with chemical, biological, or nuclear warheads, at targets many thousands of miles away.

Despite Bull's obsession with weaponry, Larry Curtis, for one, does not believe that the Canadian ballistics scientist was mentally unbalanced. "Not crazy," he said emphatically, "but very arrogant. My impression was that the guy really thought he should have been born a king. He believed that because of his superior intellect he was somehow above the law. I'm not talking about the healthy frontier antiauthoritarianism, the distaste for government bureaucracy that you'll find here in the Northeast Kingdom and throughout the North Country. Gerald Bull was a *real* outlaw. When I finally documented an airtight case against him, he simply could not believe that it had happened. I'll show you what I mean."

Larry stepped into his den and returned with a baseball cap. Instead of a team insignia, it bore the inscription BACKWOODS AGENT. The cap had been presented to him by his colleagues after Bull, indicted for arms smuggling and sent to jail for six months, was quoted in the *Boston Globe* as saying, "Some backwoods Vermont customs agents want me to be dead and want Space Research to be dead."

But why the light sentence, I asked. Surely the man who had engineered the largest arms smuggling scam in the history of this country should have spent more than six months in jail? Larry Curtis smiled that wry, ironical smile I was coming to expect in any conversation with a North Country native involving the government. "The night before the grand jury indictments were scheduled to be handed down, I got a phone call from one of the assistant federal attorneys prosecuting the case. It had been my understanding, and his, that in view of my testimony the grand jury was going to indict fifteen individuals, five companies, and three countries. But the attorney who called told me he'd gotten word that very evening from his boss that, presumably as the result of 'a telephone call from the White House,' the Justice Department intended to accept Gerald Bull's offer to plea bargain, give him and one other Space Research employee token sentences, and drop *all* the additional charges. Of course it was a coverup, to protect the government from its involvement in the smuggling."

Although Larry Curtis was nominated for Treasury Agent of the Year, he never did get over the frustration of dealing with corrupt government officials and conflicting bureaucratic mandates. Now a zoning administrator and selectman in the most beautiful village on the American-Canadian border, this tough, straight-talking agent finally got so fed up with what he describes as "bureaucratic bullshit" that he retired from the Customs Service a year early.

Gerald Bull? After waltzing out of his minimum-security institution after six months, he set up shop as an arms dealer in Belgium and later sold his Supergun technology to Saddam Hussein, who was in the process of erecting two of the futuristic weapons when U.N. forces defeated his army in the Persian Gulf War. Bull, in the meantime, had continued his work on a new scheme known as Project Babylon, designed to provide Iraq with a still more enormous gun — until one night his body was discovered slumped in the hallway outside his Brussels apartment, with two bullets from a 7.65-millimeter pistol in his head. Most probably, Larry Curtis told me, the man who believed he was above the law was assassinated by the Mossad, whose repeated warnings to stop trafficking with Iraq Bull had ignored.

In a strange postscript to this strange tale, the huge prototype of the Supergun still lies just a few hundred feet from the Vermont border on the Quebec side of Gerald Bull's abandoned compound, now growing back fast to brush and woods, like the abandoned Vermont farm fields nearby — surely the most bizarre artifact, from what is quite possibly the most bizarre story, along the entire length of the border or anywhere else in the North Country.

Notes from the False Forty-Fifth Parallel

On a map, this is one of three or four straight lines in the entire length of the border. It is an arbitrary 155-mile section, neatly trimming Quebec and touching the states of New Hampshire, Vermont and Quebec. . . . Officially this is called the 45th parallel [midway between the Equator and the North Pole] but the international boundary never really touches the 45th degree of north latitude. Hence, it is known as the false 45th.

— Marion Botsford Fraser, *Walking the Line: Travels along the Canadian/American Border*

NOON. On to the west. Phillis is vacationing with our daughter, Annie, on the coast, so there's no good reason to drop by my house in Vermont and, in order to maintain the gathering momentum of my journey, every good reason not to. This straight-as-a-string stretch of border more or less coincides with the forty-fifth parallel from northeastern Vermont to Massena, New York, and the St. Lawrence River.

Near East Berkshire, Vermont, in 1870 a handful of Vermonters were happy to urge on a contingent of insurrectionists belonging to the Brotherhood of Fenians, a tatterdemalion band of several thousand Irish expatriates who launched attacks on Canada from New Brunswick to the Midwest. "On to Canada" was their slogan; but north of East Berkshire they didn't get on very far — approximately an eighth of a mile across the border, in fact, before a small group of irate local Canadians put them to rout.

2:00 P.M. Knowlton, Quebec — an enclave of English-speaking Loyalist descendants just north of the border, where local historian Marion Phelps shows me a lively partisan monograph she has written on Vermont's "notorious" Ethan Allen, who, after the Revolution, appropriated a large chunk of nearby border territory and drove out any British sympathizers. I especially relish the title of her article: "Refugee Loyalist John Griggs. Victim of Lawlessness." At least in *this* neck of the woods, I suggest, this wasn't always the friendliest border in the world, and Marion Phelps agrees.

2:30 P.M. East Stanbridge, Quebec, in Loyalist country. The local historical society's quarterly publication still refers to the revolting American colonists as "rebels" and the descendants of Tory immigrants as "good United Empire Loyalist stock." Wistfully (I think), a member of the society, of U.E.L. stock herself, tells me, "With the recent move toward sovereignty of the Parti Québecois, my English-speaking neighbors and I are beginning to wish our ancestors had never left New England." That reminds me of something else my editor friend, Chris Braithwaite, told me just before I left home: "The variable between English-speaking Canadians and Americans is mainly political and even then it's quite subtle. In a way, though, it's as simple as this. In northern New England, one town right after another was named for Revolutionary War heroes, including the town where you live, Irasburg [named for one of those "notorious" Allen boys — General Ira Allen]. In English-speaking Canada, where I was raised, the towns are named for people who fled the Revolution. Most of my former neighbors back in Ontario like to think they'd have sided with the Loyalists. I like to think that if I'd been living in the Thirteen Colonies in 1775, I'd have been for the Revolution. Just after I moved to the U.S. in 1970, Prime Minister Trudeau declared virtual martial law in Canada. Can you even begin to imagine that happening in the United States? Unthinkable! But most Canadians seemed to think it was the right thing to do."

3:00 P.M. As if to underscore what Chris had told me, here's a Canadian customs official near the Quebec-Vermont border: "What Canadians probably value most is law and order and stability. What Americans value most is independence and personal freedom."

* * *

3:30 P.M. St. Albans, Vermont. In this Lake Champlain village in 1864 Lieutenant Bennett Young and twenty-two Confederate raiders, armed to the teeth and yelling Rebel battle cries, rode south over the Vermont border from Montreal, robbed three banks, set several buildings ablaze with a highly volatile compound of liquid phosphorus called Greek fire, which burst into flames upon contact with the air, and went hooting hell-for-leather back over the line with more than $200,000 in stolen cash in their saddlebags — leaving one bystander dead, the town in shambles, and all Vermont in a frenzied uproar after the only Civil War action on New England soil.

4:30 P.M. Northern New York's frontier with Quebec and Ontario. To cap off my jaunt along this very lively stretch of the border, an American customs official tells me that during the past decade he's subdued and arrested a fugitive with a devil's head tattooed on his chest, wanted for murdering a family of four with a pair of scissors; rescued a thinly clad illegal alien from Greece, lost in a blizzard in the densely wooded northern foothills of the Adirondacks; apprehended a sweet-talking blond beauty in the process of smuggling three international terrorists into the country; and intercepted a vanload of illegally obtained reptiles on the endangered species list, destined for Montreal. "Watch out for crates marked in red," he was warned. "Those contain venomous serpents!"

On the St. Lawrence with Dr. Solomon Cook

AKWESASNE MOHAWK TERRITORY
NO IRS FBI NEW YORK STATE POLICE
NEW YORK STATE TAX DEPARTMENT REPRESENTATIVES
OR
COUNTY DEPUTY SHERIFFS
ALLOWED WITHIN

Its bright red letters standing out sharply in the clear, late-afternoon light of northern New York, the sign loomed up ahead of me, just off the shoulder of Route 37, a mile south of the St. Lawrence. I knew that for years there had been bad blood between local Mohawk Indians and outside authorities, but even so I was surprised by the message. Here was frontier individualism racheted up to a formal declaration of political independence comparable to the Indian Stream Rebellion of a century and a half ago! If the Mohawks were serious — as I believed they were — these sentiments were much closer to those of the Québecois Separatists than the de facto independence of, say, the residents of Maine's Madawaska Republic or Vermont's Northeast Kingdom. "Akwesasne Mohawk *Territory*" — this was a geographical and political imperative, underscored by a plaque just down the highway at the Mohawk Museum, describing the Akwesasne territory (*Akwesasne* means "Land of the Drumming Partridge") as "a *sovereign nation* between the U.S. and Canada."

Like Gerald Bull's abandoned weapons compound, the Akwesasne reserve straddles the boundary. It is the only international reservation on the American-Canadian border, and its unique geographical location, enhanced by the adjacent St. Lawrence River, has created what yet another local border official I spoke with described as a smugglers' paradise. He said that as much booze flows north to Canada across the northern New York border today as came south over the same line during Prohibition, and that cigarette smuggling is more lucrative still. Three principal Mohawk entrepreneurs control the local smuggling, hiring young men, some not yet out of high school, to run hundreds of cases of cigarettes a night across the St. Lawrence to Cornwall, Ontario, on the north shore. "With the terrifically high Canadian taxes on cigarettes, a smuggler can double, triple, or even quadruple his initial investment," the official told me. "One guy here on the reservation is rumored to have made up to a million dollars a day at it."

In fact, there seems to be an unwritten government agreement to maintain a hands-off policy toward the Akwesasne Territory. "The Mohawks themselves don't regard cigarette running as smuggling," my friend told me. "Of course, they don't acknowledge that the border exists, either; they've actually torn up the local monuments marking it. The person you should see about all this, though, is Dr. Solomon Cook. He lives out on the reservation, and I can guarantee that you'll enjoy meeting him."

At seventy-five, Dr. Solomon Cook was a stocky, rugged-looking full Mohawk, with a very kind face, and shrewd, amused eyes that took in everything at a glance, then contemplated it at leisure. A horticulturalist with a Ph.D. from Cornell University, Dr. Cook was picking late raspberries behind his farmhouse at the junction of the Raquette and St. Lawrence rivers when I stopped by that evening. He invited me to sit down on a bench by his barn, and seemed glad to have a visitor. When I told him that I'd been following the "smuggling path" along the border from eastern Vermont to the St. Lawrence, he shook his head and said that these days, everyone on the reservation seemed to be looking for easy money from smuggling or from the proliferating casinos. Then he laughed and said that come to think of it, this wasn't anything new. The previous owner of his house had been the biggest Prohibition-era bootlegger in northern New York — though in those days many Mohawks also made a lucrative living as steelworkers on bridges and skyscrapers.

When I asked Dr. Cook if it was true that Mohawk steelworkers had no fear of heights, he laughed again and said no, when you're standing on a six-inch-wide beam swinging a sledgehammer two hundred feet up in thin air, you'd damn well *better* be afraid of heights. "What Mohawk steelworkers did have and to this day still do," Dr. Cook said, "was an unusual ability to concentrate on what they were doing. Probably they developed that during their early training in hunting and tracking from their fathers and uncles."

This reminded Dr. Cook of a story. Many years ago, when he was sitting for his Ph.D. orals at Cornell, a crew of North Country Mohawks happened to be framing a new high-rise dormitory just across the college quadrangle. Pointing out the window at the Indians, a professor began the doctoral examination by saying, in a half-jocular tone, "What I'd like to know, Solomon, is why you aren't out there putting up steel with your relatives?"

The question had blatant racist overtones, and Solomon Cook knew it. His reply, however, was as wise as the professor's remark was ignorant. "That's dangerous work," he told the examining committee. "Frankly, I want to live a long, long time. There's a great deal that I hope to accomplish, for myself and for my people."

Dr. Cook told me that in 1920, when he was born to Mohawk-speaking parents on a farm less than a mile from where we were talking, his maternal grandfather, a self-taught biblical scholar, suggested that the new baby boy be named Solomon. "I don't know if it was my name that did it or not, but I made up my mind early on that I was going to get a first-rate education," he said. "Later I had an inspiring high school teacher who told me I could do anything I put my mind to. I put my mind to being one of the first Mohawks to leave here and acquire a college education."

After a while Dr. Cook and I walked out in back of his barn, through more raspberry bushes, into a long rectangular pasture leading down to the river. "My ancestors picked a beautiful spot to settle in," he said. "Today, though, the big industries upriver have poisoned miles of the river, and there isn't a single farm of any kind on the reservation, though we still turn out the best sweetgrass baskets and ash lacrosse sticks in the world. This is sweetgrass, by the way."

Dr. Cook bent over and pulled up a long, thin, rushlike blade with a delicate fragrance.

After college, the navy, and graduate school, Solomon Cook put in a stint teaching agriculture courses at Cornell. He didn't care for academic life, though. "You know, for centuries Indians have been stereotyped as wild and crazy drinkers. But there was at least as much hard drinking at the weekend faculty cocktail parties as you'd see up here on the reservation. If you wanted to climb the academic ladder, you had to go to those parties and play ball with the boys. Well, I was too independent for that. After a year of it I came back to teach vocational agriculture at the local high school and run my own little farm, and I never regretted it once. I felt emancipated up here in a way I didn't at Cornell. I didn't need to worry about cocktail parties and playing games. I was free to teach, which, you know, is a noble calling."

As we admired the St. Lawrence, now turning a deep purple in the sunset, Dr. Cook said that during the early 1980s he served a three-year term as chief of New York's St. Regis Mohawks. It was a time of great turmoil for the tribe, he explained, with bitter internal fighting between two Mohawk factions and even greater friction between the tribe as a whole and members of the local white community. The reservation was in an uproar, with state police barricades on the highway nearby, hostile rhetoric at the tribal meetings, even shootings and bombings. The local district attorney had indicted scores of Mohawks on charges ranging from disturbing the peace to assaulting police officers. Dr. Cook had grave doubts about undertaking the chief's job during all of this strife, which amounted to "an undeclared state of war."

"Ultimately, though, I felt I owed it to the tribe to give the job a shot," Dr. Cook told me. "I had three objectives. I wanted to disband a dictatorial Mohawk police force, which was actually going around to people's homes and beating them up, like the Gestapo. I wanted to balance the tribal council by reintroducing some traditionalists. And most of all I wanted to heal the wounds and bring permanent peace back to the community. Well, I can tell you, Mr. Mosher, it was a terrible time for me and my family. My house was bombed twice. All the windows in my shop were broken. My life was threatened. I hated to start up my car in the morning for fear of what might happen. But with my wife's support I served out my term and for the most part I achieved my objectives. I even persuaded the local D.A. to drop nearly all of the charges against tribal members, by arguing that most of our Mohawks had never been in trouble with the law before and very probably never would be again. Today we're fighting General Motors and Alcoa, trying to get them to clean up their toxic waste sites by the river. But we're using lawyers, including Mohawk lawyers, not bombs. Now I want to show you one more thing before dark."

Dr. Cook led me down through the twilit fields behind his house to the bank of the St. Lawrence, where he gestured down the river. "See that tongue of land over there? Just north of where the Raquette comes in? And the lights of the superfreighter steaming upriver just beyond it?"

I nodded.

"That point of land is the first United States territory that ships like the freighter pass coming up the seaway from the ocean. It's Mohawk land, and it would be an ideal place for a seaport. For many, many years I've urged the tribe to consider building one there, like the seaport upriver at Massena."

Pointing out over this great, muscular North American river marking the border, Dr. Cook looked more like one of the famous Mohawk chiefs of the eighteenth or nineteenth century than a modern-day scholar with an Ivy League doctorate. Being here with him as he articulated his vision for his tribe was one of the best moments of my trip.

Then he smiled his kind smile and shook his head. "Casinos and cigarette smuggling bring in quick money, easy money. But an active seaport — now that would be a guarantee of legitimate jobs and prosperity for a long, long time to come. That's *my* dream for the Land of the Drumming Partridge."

The Farm

Some good small farmers remain, and their farms stand out in the landscape like jewels. But they are few and far between, and they are getting fewer every year.

— Wendell Berry, "Nature as Measure,"
in *What Are People For?*

That night at my motel on the St. Lawrence, within earshot of the roaring 200-horse outboard engines of the Mohawk cigarette runners, I thought back over the first leg of my trip and recalled something that Ti René had told me: the border country was a hard place to make a living. Yet as René had also said, it was a beautiful place and a good place to live, particularly for individualists like him and Harry Hughey and Solomon Cook and, for that matter, myself — people who, whatever their professions, needed above all else the elbowroom to be free, to run their own show, to do things their way.

The first North Country individualist I ever knew was my grandfather Mosher. After I left Chichester, when I was ten, my boyhood was transformed by my moving with my parents to his upstate New York farm, not far from Solomon Cook's, actually, as the crow flies. We lived on the Farm, as we called it, with a vast, extended family, which transcended all the borders of what is usually thought of as family to include a wondrous gallery of hired men, housekeepers, distant cousins, and great aunts and uncles, many of whose blood connections to my family were so obscure and tenuous that even my genealogy-minded grandmother couldn't trace them with certainty.

Now, decades later, winding up the first part of my journey in a Canadian-looking countryside of endangered forests and towns and railroads and farms, it occurred to me that growing up in the North Country had been a process of encountering one inevitable major change after another. My whole subsequent career had, in some ways, been an attempt to understand those changes, both by preserving in my fiction the North Country I'd known as a child and by exploring critical periods of change in it and in my own life. As I looked out over the big dark river at the lights of Cornwall, Ontario, I found myself thinking of yet another long-ago fall day of almost cataclysmic change in my boyhood, much like the day when I had walked up to Uncle Reg Bennett's house through the dying village of Chichester.

I was fifteen years old and still living with my grandparents and that marvelous assortment of relatives in the dilapidated Mosher farmhouse near Lake Ontario and the Canadian border. It was a crisp, sunny Saturday morning in early September. My grandfather had gotten up at his usual leisurely hour, put on a clean white shirt and a tie, as he had every day of his life since he was twenty-one years old and first elected justice of the peace, and eaten a big breakfast in the farmhouse kitchen. Here in this comfortable room over the years, I'd sat quietly on the corner woodbox and watched my grandfather dispense not only justice but advice, concern, wisdom, cash (when he had any), and time — most of all time, since Howard Leroy Mosher had time for everyone — to the hundreds upon hundreds of farmers and businessmen and lawyers and young couples and people in legal trouble and every other conceivable kind of trouble, who came from near and far to seek his counsel.

"Howard Frank, I want you to take a walk with me this morning," my grandfather said to me as he finished his breakfast.

Though Gramp often asked me to accompany him on his daylong automobile forays into northern New York's hinterlands in search of general stores selling extra-sharp cheddar cheese off the wheel, or antiques for my grandmother, or the best homemade blueberry pie in the North Country, or some especially outlandish character he'd heard about and wanted to meet, he rarely walked anywhere he could drive or be driven. I honestly think he didn't want to get his thrice-polished dress shoes dusty. Where would we walk today, I wondered, and for what purpose?

Sometime in the middle of the morning, we started down the lane behind the farmhouse. My grandfather, in his midsixties but well preserved, a small man (nearly a foot shorter than my six-foot-tall father), with thick white hair parted in the middle, humorous blue eyes, and clear skin unroughened by outdoor exposure, walked slowly and deliberately. On our left we passed the hunchbacked green trailer where Old Bill, the last in a long cavalcade of hired men, lived. We walked past the horse barn, still in good condition, where my grandfather had once kept six working Clydesdales and a pair of Morgan driving horses; past the gigantic three-story cow and hay barn; and on down the lane to the brightening fall woods at the end of the fields. Then my grandfather began to speak, in that easy, speculative, spellbinding manner that people came from twenty and fifty and one hundred miles away to listen to and that I had listened to all my life.

Gramp began by telling me stories of his childhood. He talked about his mother, who'd come alone to the States from Ireland when she was fifteen — just my age — landed a job at my great-great-grandfather's house as a hired girl, and married my great-grandfather Mosher. He spoke of his work as a criminal investigator for the local district attorney and his particular expertise in finding missing persons, which had taken him all over the continent, and of his lifelong interest in acquiring the stories of unusual individuals of all kinds, from railroad tramps and carnival roustabouts to migrant apple pickers and Indians from the nearby reservation, many of whom still periodically stopped by the Farm for help. And he spoke frankly of his love and admiration for my grandmother.

On my grandfather yarned, telling me how he'd bought the Farm for my grandmother as a wedding present when he was just twenty years old and a successful real estate dealer. "'The smartest young man in New York's North Country,'" my grandfather said to me. "That's what people called me, Howard Frank. But there was one thing the smartest young man in the North Country wasn't smart about, one thing he couldn't foresee."

"What was that, Gramp?"

"The Depression."

"You made out all right." I gestured back toward the hayfields, cut just last week for the third time that year, toward the barns and the brick house.

My grandfather smiled. Then without a shred of self-pity he said, "I'd have made out better if I'd gone to college and law school like my brothers. With a law degree, I could have been anything. I could have been governor of the State of New York, like my friend Thomas E. Dewey."

My grandfather wasn't boasting. I doubt that Howard Leroy Mosher ever uttered a boastful word in his life. What he said was simply true. I believed this then and believe it now.

"Smart is one thing, educated is something else," my grandfather said. "Sure, I was a bright young man. I just wasn't quite bright enough to see that to get beyond a certain point in politics, I needed that degree.

"The fact of the matter is that hardly anyone can make a living farming in the North Country anymore," he went on. "But all this land is well situated and has the best drainage within twenty miles. It's not worth much for raising crops, but it's worth a great deal to build houses on. I'm in the process of selling it to a real estate company, all but the five acres the house and barns sit on, and the income from the sale is going to provide first-class educations for you and my other grandchildren. That's your inheritance."

Sell the Farm! This was unthinkable. I had always assumed that the Farm would remain in the Mosher family in perpetuity. I'd own it myself someday, live there with a family of my own, and write stories about growing up with my grandparents. I started to protest. We'd all get our education anyway. I was going to be a writer. As soon as I got out of high school, I planned to hitchhike across the country, coast to coast, like Jack Kerouac, and get my education that way.

But the smartest man in New York State's North Country just smiled and said, "The Farm will pay for your education,

Howard Frank. That's why I've hung on to it all these years. The stories can be my special legacy to you. But the Farm is your inheritance, and I want your inheritance to be your education. I don't want to think you'll ever have to tell your grandchildren that you could have been or done anything, except for the lack of a degree."

Part Two

THE NORTH COAST

The Fisherman and the Pipe Carrier

I never can play on de Hudson Bay
Or mountain dat lie between
But I meet heem singin' hees lonely way
De happies' man I know.

— William Henry Dummond, *The Voyageur*

They came from the cobbled back streets of Québec City, from Trois-Rivières and Tadoussac and the remote farming communities of Quebec's Eastern Townships. Each spring, immediately after ice-out, they gathered near Montreal to set out up the St. Lawrence River in freighter canoes thirty-five feet long, and for the next six months they paddled sixteen, eighteen, sometimes twenty hours a day to reach the fur trading posts beyond the Great Lakes and return to Montreal before the dreaded winter trapped them in the bush. They were the fabled voyageurs of the Hudson's Bay and North West trading companies, men I'd first heard of from my mother's mother, who had read me Henry William Drummond's marvelous long dialect poem, *The Voyageur*; and for nearly two centuries they were a breed of men apart from all others on earth.

What a spectacle their flotillas must have made as they headed upriver in brigades of twenty to thirty canoes, the men as colorful as songbirds in their gaudy wool shirts and wide crimson sashes and feathered hats. Their flashing paddles were tipped with vermilion, and the river rang from bank to bank with their ancient French rondels, some with as many as one hundred and eighty verses. The voyageurs themselves were on the small side, wiry and compact, which saved room for trading goods in the outward-bound canoes and, more important, for the precious beaver pelts they would return with from the frontier. But these men were incredibly tough and resilient, capable of paddling against powerful currents from first light until long after dark, bucking oceanic whitecaps, and racing at a flat-out run across swampy portages with two-hundred-pound packs on their backs. Some drowned in the rapids and in Great Lakes tempests. Others collapsed from heart failure or strangulated hernias or were literally blinded from the bites of mosquitoes and black flies that rose out of the swamps to attack them in clouds. A few of the voyageurs, unable to stand up to the grueling pace, went bush crazy and rushed into the wilderness, tearing off their clothes and shrieking like wild animals, never to be seen again.

Hurrying, hurrying, they were forever hurrying to get to Ottawa and Lake Nipissing and Sault Ste. Marie, then on across Lake Superior to the Pigeon and Rainy rivers and the wild and mysterious Lake of the Woods. For much of their path they followed what is sometimes called America's North Coast — the U.S. frontier along Lake Superior and beyond. Much of their route coincides with the present-day border, and much of it is nearly as gorgeous today as it was when the first voyageurs set eyes on it.

With a few side excursions, this is the route I chose for the next thousand miles of my trip.

I drove into Sault Ste. Marie, Ontario, directly across the St. Mary's River from Sault St. Marie, Michigan, in the middle of the afternoon, having comfortably covered, in less than a single day, the trek from Montreal that took the voyageurs five backbreaking weeks. These sister cities facing each other across the largest lock in the world are among the small handful from Maine to the Pacific that share a name. It was conferred by the Jesuit explorer Father Marquette, who, legend tells us, was vouchsafed a rare glimpse of the Virgin Mary hovering — for exactly what purpose we know not; I like to think she was fishing — in the mist over the thundering rapids or "sault" connecting Lake Superior and Lake Huron. In any event, the entire region for miles around has long been popularly known to Canadians and Americans alike simply as the Soo.

At first the boardwalk along Sault Ste. Marie's riverfront seemed deserted that afternoon. Then I saw a lone spinning rod leaning against the railing, its thin monofilament line angling far down into the river current where it set in close to the Canadian shore. I looked around for the fisherman and spotted two beat-up sneakers sticking out of the shrubbery a few feet away.

"How's the fishing?" I called out.

The sneakers shifted, there was a rustling, and a rugged young man in a painter's cap and white pants spattered with green and yellow emerged from the bushes, where he'd been sitting out of the wind, waiting for a bite. Bob Bagnall told me that he was a house painter by trade and a fisherman every minute that he wasn't painting houses. Right here, within a hundred yards of the border and just across the river from the busiest shipping channel in the world, he regularly caught more than a dozen different kinds of game fish.

Two fishermen meeting for the first time rarely need more than five seconds to break the ice. No sooner were those words out of Bob's mouth — a *dozen* different species of game fish — than we were off and running on our mutually favorite topic, fishing and everything having to do with it.

Bob, who was in his midtwenties, told me that he'd fished the St. Mary's here below the site of the first lock built on the river — a wooden structure completed in 1798 to enable his voyageur ancestors to reach Lake Superior with their freighter canoes — almost daily, summer, fall, winter, and spring, since he was a small boy. Like Tom Sawyer, he'd sneak out his bedroom window after the rest of the household was asleep, drop off the porch roof, and slip down to the river for a little clandestine nocturnal angling for lunker brown trout and catfish. In fact he'd fished here so long that he could usually tell what sort of fish he had on just by the way it was biting. "A pickerel hits very lightly, eh? They jab, jab, jab. A trout of any kind, he'll strike with much faster, stronger jerks. Bass, both largemouths and smallmouths, they'll run with the bait, stop, turn it in their mouth, run again. After they stop the second time, that's when you set the hook. Bottom feeders, sturgeon, catfish, whitefish, they suck in the bait and give a slow steady pull. But when you see your rod tip bend right down hard and the line starts smoking off the reel, that's very likely a salmon. Then, mister, you've got a battle on your hands!"

Bob told me that he routinely caught four different kinds of trout off the boardwalk. Besides brown trout, rainbow trout, and lakers, he quite frequently latched onto brook trout up to three pounds. Brook trout! Harry Hughey's trout, the denizen of the Laurentian waters my father and uncle and I fished when I was a boy and the Labrador rivers I'd fished with my own son, and of all North American game fish the one that requires the coldest and purest fresh water to live and spawn in. Brook trout cannot live for long in polluted lakes and rivers, and I was frankly astonished when, a few minutes later, Bob hooked, played, netted, and released one of about a pound and a half.

Bob had devised an ingenious method for netting his fish. A conventional long-handled net would have been of no use here on Sault Ste. Marie's boardwalk, which ran along the top of a stone quay a good twenty-five feet above the river. Nor could you ever crank a ten-pound salmon, or even a five-pound trout, up to the railing; the lightweight monofilament line Bob used would snap like thread. To deal with this problem, he had constructed a homemade net from a spokeless bicycle-wheel rim attached to a stout, neon-orange nylon rope thirty feet long. With a bit of help from his feet to pay out slack, he played a fish off his rod with one hand, lowered the net into the water with the other, maneuvered the fish over it, then lifted the net up out of the river with his catch inside.

An active conservationist, Bob frequently volunteers to assist with federal and provincial stocking and fish-counting efforts in the Soo area. Still, he was quick to point out that like the cod on the Grand Banks off the Atlantic coast, the populations of whitefish and lake trout in the Great Lakes border waters have been so badly devastated by centuries of overfishing and pollution (not to mention the ravages of the sea lampreys, which attach themselves to the sides of game fish and suck out their life blood like vampires) that commercial fishing in Superior and Huron is nearly a phenomenon of the past. "I myself kill very few fish these days,"

Bob said. "Once in a great while I'll keep a salmon or a trout to give to a relative or a friend, and over the years I've kept one trophy fish of each species to have mounted. All the rest go back in the river. That's where they belong, eh? If sport fishing in the Soo is going to have a future."

Bob Bagnall and I shook hands; I wished him luck and headed up the boardwalk toward my car. When I glanced back to wave, he had faded into the shrubbery. The only trace of his presence was his black spinning rod leaning against the railing at an expectant angle, as Bob waited out of sight to come forth and battle great shimmering emerald pickerel, pike as long as your leg, trout as bright as autumn hills, dark fierce muskellunge, prehistoric-looking sturgeon, and four kinds of salmon, which he lifts up to admire in his bicycle-rim landing net, then releases unharmed to swim free in the St. Mary's River, where they belong.

I'd never ridden in a limo before, and truth to tell, I didn't much like riding in this one. For one thing, the driver and I were all alone in a vehicle designed for a dozen passengers. For another, I couldn't shake the macabre foreboding that I was heading to a funeral, my own for all I knew, in a hopped-up white hearse. In fact the limo was one of a fleet belonging to the American Sault St. Marie's local Sioux Indian tribe, which also owned and operated the gambling casino whose stadium-sized parking lot, jammed with cars and tour buses, I was now silently, magisterially, and, I felt, very foolishly rolling into.

I was flabbergasted by the size of the place. The casino was as big as a major-league hockey rink; and at ten-thirty this weekday evening in late August it was packed, humming with the color and noise and activity of round-the-clock poker, keno, craps, slots — a mini–Las Vegas in Michigan's remote Upper Peninsula.

"Howard? Howard Mosher?"

My heart sank. Evidently I was about to be waylaid by some acquaintance from heaven knew where, just as I was ready to plunge into the bustling casino for a firsthand look at one of the oddest, if most lucrative, businesses in America's North Country today. To my relief the man who had called out my name wasn't an acquaintance at all but a stranger of about thirty, who introduced himself as Allard Teeple, a Chippewa pipe carrier and an adviser to the local Sioux chief. He'd be glad to tell me anything I wanted to know about the casino, the Sioux tribe, or the Soo area, he said with the sort of unwary friendliness that visitors are so much more apt to encounter in the Midwest and West than in New England. But how had Allard Teeple known who I was? From the motel proprietor who called the limo for me? I decided to leave it a mystery.

We sat at a table in the casino's restaurant, where Allard began our conversation with an account of his hand-to-mouth boyhood on the Upper Peninsula. He had been raised by Chippewa elders after both his parents died. Several years ago the elders had offered him the tribal pipe, symbolizing spiritual leadership. Allard declined it, feeling that he was too young to accept the responsibilities accompanying the honor. During the next year or so he was offered the pipe twice more. The third time, he told me, it simply showed up out of the blue at the place where he happened to be staying, hundreds of miles from home. This time, beleaguered by dreams full of portentous symbols, he reluctantly accepted it. It was a tough choice. But once he'd made it, he was almost immediately faced with a harder decision still.

"At that time the Upper Peninsula tribes were debating the pros and cons of following the lead of other tribes that had set up gaming operations. They wanted my opinion on the matter, and to tell you the truth, I had serious concerns about the entire concept of gaming. Finally I went on a retreat to contemplate the question. I walked far out into the prairie to a very wild place that I thought was known only to Indians and fasted there for three days and three nights. By the fourth day, when I still hadn't received an answer, I stopped drinking water and restricted myself to walking in a tight circle.

"Within a few hours I'd worn a shallow dirt path into the grass. Some time later, how long I'm not sure, I felt something turn under my bare foot. I glanced down and saw the sun glint off it, the way it would off a shard of bottle glass. I knew an Indian never would have brought such a thing to a sacred place. 'Well, great,' I thought. 'No matter where we go we can't get away from them.' But when I reached down and picked it up and brushed off the dirt, I saw that it was a square green die, with numbers written on the sides, like a gambling die. That green die was my answer. Now I knew what to advise my people to do."

"To open a casino?"

"Yes. But more than that, I knew somehow that gaming would be a steppingstone to independence."

I nodded toward the casino floor. "Has it been?"

"Of our current employees," Allard Teeple said, "twelve hundred came directly off the welfare roles. Many became financially independent for the first time in their lives. Still, nobody in the tribe wants to be entirely dependent on gaming. So far, we've invested our casino profits into nineteen other businesses. We rebuild core steering units for front-wheel-drive cars, and we own and operate a neon sign shop and an Anheuser-Busch branch plant. We've started a financing company, which put up the local shopping mall, among other ventures, and we've just opened a surgical instrument factory. Out of all the social and economic problems in Michigan's Upper Peninsula today, unemployment has to be the worst. Right now we're the single biggest employer in the area. Bigger than the paper companies and bigger than the mines. These days, when we go into the stores downtown, people treat us with respect. And, frankly, to a Native American, that's something you can't begin to put a price tag on."

Notes from the Upper Peninsula

Murders must happen some place, of course, and this one and the subsequent trial took place on the water-hemmed Upper Peninsula of Michigan, simply U.P. to its inhabitants. The U.P. is a wild, harsh and broken land . . . a jumble of swamps and hills and endless waterways . . . perhaps more nearly allied with Canada by climatic and geological affinity.

— Robert Traver, *Anatomy of a Murder*

There was nothing but the pine plain ahead of him, until the far blue hills that marked the Lake Superior height of land.

— Ernest Hemingway, "Big Two-Hearted River"

8:00 A.M. Within an hour after leaving the Soo this morning, I drive out from under the lake-effect morning cloud cover into a sunny, summery day. Still, at this time of year you can't travel ten miles anywhere in the North Country without some reminder of impending winter. Back in Maine it was highway crews stringing up snow fencing. Here on Michigan's Upper Peninsula, signs announcing ICY PAVEMENT and BLOWING SNOW keep cropping up. Otherwise, the countryside is wooded and poor-looking, with rusted trailers and tarpaper shacks hunkered back off the road in the scrubby underbrush, reminding me of northern New England.

9:00 A.M. I am looking down into the tannin-stained depths of a stream in Seney, the starting point for young Nick Adams in my all-time favorite fishing story, Ernest Hemingway's "Big Two-Hearted River." Seney was a train trestle over a trout stream when Nick visited it in the story. The entire town, including all thirteen of its saloons, had burned to the ground around the turn of the century, been rebuilt, and promptly burned to flinders again. Today it's still scarcely more than a whistle stop, just one store and a few houses, and the stream still looks to be a likely spot for the big speckled beauties that Nick admired off the trestle. I'm tempted, again, to fish but decide not to. There'll be plenty of time for fishing in Montana.

9:30 A.M. The back of a loggers' boot hill just outside Seney, a hodgepodge of wooden markers tilting crazily above sunken graves overrun with myrtle and ground pine and inscribed with the following legends:

CHAS DEWEY AGE 33 DIED FIGHTEN.

UNKNOWN. DIED IN A BAD FIGHT.

NO NAME. KILLED ON THE RIVER.

10:00 A.M. North of town a few miles I stop again and walk a short distance off the road into the woods. Suddenly I could be standing next to Nick Adams, looking out over the dark islands of pines in the sandy barrens at a faraway, sparkling river, and the distant blue ridge just south of Lake Superior, exactly as Hemingway described them. But I resist the temptation, a few minutes later, to strike off onto a side road to visit the actual Big Two-Hearted, knowing that the only place I'll ever truly find Nick's river is on the pages of Hemingway's eternally lovely story.

10:30 A.M. PRIMITIVE ROAD FOR 50 MILES. The sign intrigues me. Just how primitive is a primitive road? Do you need a four-wheel-drive vehicle to attempt it? I pull over to reconnoiter. Beyond the tantalizing sign, the road snakes between Lake Superior and a dune as high as a Vermont hay barn. In spots sand has drifted hubcap-deep over the road. This is as close to the Upper Peninsula's original wilderness as I'll ever get, I think, and pull out onto the narrow track to see what this primitive road is all about.

11:45 A.M. High atop a sand cliff at the west end of the Grand Sable Banks, in the Superior National Forest, I come upon the site of what, in its day, must have been as original a logging enterprise as any in the North Country. On this spot a century ago, white pine logs hurtled down a five-hundred-foot wooden chute resembling a ski jump from the clifftop to the lake below, to be rafted east to the big sawmills in Grand Marais, Whitefish Point, and Paradise. The half-ton pine logs sped lakeward so fast that their friction ignited flash fires on the smooth wooden surface of the chute; men stationed at intervals on the face of the cliff doused the fires with buckets of sand. Yes, I know all too well what havoc those red-shirted woodcutters wreaked upon the U.P. forests, and I find myself rejoicing that at least theoretically such devastation can never again spoil this or any other national forest. But even so, the thundering log chute is a stirring sight to conjure up in this wild, empty, and, once again, after nearly two centuries of rapine, unspoiled place on Michigan's North Coast.

5:30 P.M. Marquette, Michigan. The waterfront is unaccustomedly quiet this afternoon because of a Mesabi miners' strike. No freight-car hoppers line the Burlington Northern's looming ore docks, which gleam darkly in the watery sunshine, thrusting out into the harbor like truncated railway trestles going nowhere. Yet even with the port shut down, Marquette has the rugged look of a working harbor. With its hulking docks and hilly, narrow streets that all seem to run down to the lake, it's a promising place to continue my search for Upper Peninsula lore.

6:00 P.M. Outside Marquette's red sandstone courthouse, the site of a famous libel trial in 1913. The flamboyant Theodore Roosevelt had brought a libel suit against an Upper Peninsula newspaper, and in the grand old tradition of San Juan Hill, the last of our two-fisted presidents won his case. For compensation, the Marquette jury that found in his favor awarded him exactly six cents: "the price of a good newspaper." Some decades later a sensational murder trial in the same courthouse inspired a brilliant young Marquette district attorney named John Voelker to write, under the pen name Robert Traver, a story set in the wilds of the U.P. and involving an alleged rape, a shooting, and some of the most exciting courtroom scenes in an American suspense thriller. Overnight it became a runaway bestseller and the basis for an enormously popular film of the same name, *Anatomy of a Murder.*

6:30 P.M. And here, to my delight, I discover the signatures of *Anatomy*'s stars — Jimmy Stewart, George C. Scott, Lee Remick — inscribed in the sidewalk outside Marquette's Nordic Theater, where the film premiered. Yet as I stand on the main street of Voelker's rough-hewn port, gazing at the famous names outside the theater, I have the sensation that something's askew. I look up and down the sidewalk, mainly empty at the weekday supper hour; look back at the signatures; and then it hits me. The cement blocks in front of the Nordic look much more recent than the rest of the sidewalk. In fact, they look far too new to be authentic. It takes an hour of inquiries, but finally the editor of *The U.P. Action Shopper News* enlightens me. Back in 1959 a number of Marquette's most influential townspeople deemed the movie *Anatomy of a Murder* risqué because of its focus on the Remick character's panties, which were trotted out and used as sensational evidence in the courtroom scenes. Soon after the sidewalk blocks with the offending signatures were cast, Marquette's scandalized town fathers came together and decreed that they must never see the light of day, at least within a country mile of the city's prim and proper main street (overlooking, I'm told, a whole row of waterfront taverns and whorehouses). Fortunately, a forward-looking local farmer rescued the inscribed slabs from the demolition ball and stashed them away in his barn, where they languished in ignominy until 1986, when they were rediscovered, brought out into the light, and installed in front of the Nordic. And so it came to pass that for twenty-seven years the delicate sensibilities of this wide-open shipping town were preserved from the Babylonian influence of Jimmy Stewart's name on Main

Street, small-town Victorianism having prevailed over small-town boosterism.

And thus, I move on west along the route of the voyageurs, deep into iron-ore country, laughing to think how tickled John Voelker must have been by this splendid instance of the blue-nosed hypocrisy he poked fun at throughout his distinguished literary and legal careers in the harsh and beautiful Upper Peninsula.

The Great Bernie Silverman

Men sometimes speak as if the study of the classics would at length make way for more modern and practical studies; but the adventurous student will always study the classics.

— Henry David Thoreau, *Walden*

It's well known that most people of my generation remember where they were and what they were doing when they first heard that President Kennedy had been shot in Dallas. (I was about to take a German test at Syracuse University.) It occurred to me now, driving west through the countryside of the U.P., that I could still recall exactly where I was when I first read Hemingway's Nick Adams stories and who had recommended them to me — the Great Bernie Silverman, a door-to-door brush salesman whom I worked for a couple of years when I was in college.

Bernie had conferred the sobriquet "Great" on himself because he could sell anything to anybody. During the summer before my sophomore year at Syracuse, he sold me on the idea of "opening up" the remote villages of New York's northern tier to what he called "the line" — no easy task, I can assure you. I'd just had an off week, selling less than a hundred dollars' worth of brushes and household appliances and scarcely making enough commission to pay for my gas. Bernie, as was his wont when one of his salesmen started to lapse, had agreed to meet me early the following Monday morning near Watertown, to give me a refresher course in selling the line. Actually he'd done this quite frequently, maybe once a month, over the past year or so. But between house visits, where he never failed to sell up a storm, he talked books to me nonstop.

How did he do it, this loquacious intellectual with a Brooklyn accent, seemingly so out of place in the tiny North Country farming communities I'd grown up in? A rotund, smallish, brusquely cheerful man, Bernie had a world of confidence; but what under the sun was his secret? If I made a sale at one out of every four or five houses, I felt I was doing well. The Great Bernie Silverman batted well over .500 while seeming to concentrate on other things — mainly the books he so dearly loved.

Bernie was interested in me and my career plans, too, since he had once dreamed of being a writer himself until, as he cheerfully and frequently announced, he discovered that he had no talent. What he clearly did have talent for (besides selling brushes) was teaching; why he hadn't become a teacher I could never divine, except, perhaps, that as a salesman and sales manager he made a great deal more money than he ever could have earned in education.

"You want to write stories, Mosher?" he said today as we hit the pavement of a forlorn little mill town where I'd once journeyed with my grandfather to buy sharp cheese. "Then you have to read stories, especially the classics. You can't read too much."

Stepping briskly up onto the porch of a weathered bungalow, he pointed his index finger at me and repeated, "Read!"

"Read what?" said a tired-looking woman with her hair up in curlers as she opened the door, at which moment I'd have bet dollars to doughnuts that even Bernie the Great wouldn't make a sale here. He did, though; before the woman in curlers had any idea what he was up to, he'd started a paper fire in her sink, put it out with a hand-held $7.50 fire extinguisher, and sold her three of the things, while she told him about a fight she'd had with her daughter's husband.

"Bernie, how do you do it?" I said on our way back to his car. "I couldn't have sold a fire extinguisher or anything else to that woman to save my life. You must have been born with a silver tongue."

"Ask me something hard, Mosher. My secret is, I do what you've got to learn to do if you want to be a writer. I listen to people. What that woman told us, about the squabble with her son-in-law? I was listening to her all the time I was filling out the sales slip."

"You mean act interested in what the customer says to you?"

Bernie smote his forehead. "No, I *don't* mean act interested in what the customer's saying. *Be* interested. Don't act. *Be.* Listen to what people say and *be* interested. So. Tell me what you're working on now."

"A fishing story. It's about this kid who wants to catch a record trout —"

"Show it to me when you get it done," Bernie said. "In the meantime, how many fishing stories have you read?"

I was momentarily stymied. Then I thought of Hemingway's *The Old Man and the Sea.*

Bernie nodded. "Got him the Nobel Prize. But you want to read a *real* fishing story by Hemingway? I think I can fix you up."

Later, in a used bookstore on a back street in Watertown (a rough-and-ready little North Country city that Marquette, Michigan, somehow reminded me of) Bernie located a secondhand copy of Hemingway's Nick Adams stories, *In Our Time.* He checked "Big Two-Hearted River" in the table of contents and handed me the book. "Read it and weep, Mosher," he said. "But at least read it."

And I did, right there in that out-of-the-way bookstore, at the end of a good day with my boss and mentor, the Great Bernie Silverman, who, when I'd asked him why he didn't write, had said good-humoredly, with irony and relish, "No talent, Mosher. No talent for anything except selling the line."

Wild Rice and Blue Rollers

I read two or three shallow books of travel in the intervals of my work, till that employment made me ashamed of myself, and I asked where it was that *I* lived.

— Henry David Thoreau, *Walden*

Wisconsin still vies with Vermont at national dairy festivals for the tangiest cheeses, but up in its northern hinterlands the next morning I passed no dairy farms or farms of any kind, just that huge inland freshwater sea, Superior, to the north, and dense woods stretching unbroken for mile after mile to the south. At Bad River, in 1850, the Reverend L. H. Wheeler convinced the United States government not to evict the local Chippewa tribe from their ancestral home on the river's wild-rice grounds. The Indians had been slated for deportation to barren scrub country to the west, and after visiting that wasteland, Wheeler bluntly told Washington that it would be kinder to shoot them. This sunny morning a century and a half later, twenty or so descendants of the people Wheeler had helped were creeping up marshy backwaters in canoes, harvesting the highly esteemed grain that their ancestors had fiercely protected, often at the cost of their lives, first from the Sioux and then from the white settlers. I walked down beside the river to watch, and a tribesman in his seventies named John Little Raven offered to take me out for a quick spin in his homemade wooden canoe.

Northern wild rice grows like cattails, standing three or four feet above the surface of the water near shore. To reap it, John Little Raven used two short wooden sticks. With one he gently bent the long wavy stalks over the canoe gunwale. Then he lightly whacked off the ripe grain kernels atop the plants with the stick in his other hand, releasing the undamaged stalks to spring back up and bear another season: a simple, effective, and ingenious method of harvesting and preserving a grain so coveted that it sells for up to twelve dollars a pound. "Even so, this is a slow way to make a few extra dollars," John said as he paddled me back to the bridge. "These days we do it mainly as a traditional community activity. Fall's a nice season to get out together, and this is something we look forward to all summer. If it was the money we were interested in, I guess we wouldn't live up here in the first place. I've got four grown sons who've all moved away to earn better money. You know what I tell them? 'You fellas live down in Milwaukee and make big salaries. That's okay. But you spend fifty weeks a year slaving for somebody else so you can spend two weeks of vacation back home doing the things you like.' Big money isn't everything. Right this minute my boys are working in a steel mill for twenty dollars an hour. But their hearts are up here in the wild-rice fields. Me, I'm where my heart is."

On the east side of Superior, Wisconsin, I passed under three elevated, boxed-in conduits leading down to a row of lakeside ore docks. At the next side street, I turned off Route 2 and headed out toward a low range of hills south of town, keeping the conduits in sight. Eventually I wound up at the Burlington Northern freight yard, where a sprawling network of tracks came in from the west. BN AUTHORIZED PERSONNEL ONLY a sign announced, but the yard gate stood invitingly ajar, the freight yard was empty, and I decided to authorize myself to enter. Out in North Dakota a few days later, this cavalier approach would nearly get me into very hot water; today, though, I breezed through the gate onto the posted property of the BN, where, so far from being evicted, I was graciously received by Wes Johnson, an amicable veteran railroad man who was happy to give me a tour of the yards.

"This is where the ore trains come in," Wes said. "They're one hundred and sixty cars long and each car can carry ninety tons. It takes more than four trains to fill a superfreighter."

The BN yards in Superior were temporarily down to a skeleton crew because of the miners' strike over in Minnesota's Mesabi Range. Ordinarily this would have been a much busier place. As we walked across several sets of tracks, inhaling the good, tarry scent of freight yards everywhere, Wes explained that Burlington Northern was an amalgamation of three lines: the Great Northern, the Northern Pacific, and the West Coast and Chicago.

Wes pointed at the three covered conduits I'd followed up from the lakeshore. "I'm a maintenance foreman on the Blue Rollers," he said. The Blue Rollers were the steel-mesh conveyer belts inside those elevated tunnels. Sixty inches wide and five miles long, the belts convey ore from the freight yard to the docks, where it's dumped into the holds of superfreighters bound for the Soo locks and the steel mills of Chicago, Cleveland, Milwaukee, and Detroit.

"Come on over to the Hump," Wes said. "Forty years ago I started work here as a switchman. I'll show you how we managed the operation back in those days."

Just below a tall scales house beside the tracks, where the ore is weighed before being poured onto the Blue Rollers, was a low knoll, known in railroad jargon as the Hump. Wes explained that before the advent of the Blue Rollers, the ore was sent down the five-mile grade to the docks in the same hopper cars that had brought it from the Mesabi. A switching engine pushed a string of hoppers onto a spur leading up and over the Hump, whereupon the cars rolled merrily downhill to the docks under their own power. As each car crested the Hump, a switchman like Wes Johnson dashed up, yanked free the knuckle pin connecting it to the hopper behind, sprang aboard, and got a free trip to the lake.

"You didn't want to forget to test the brake wheel," Wes told me. "Otherwise it could be a bobsled ride, with a nasty crackup when you hit the car ahead of you at the bottom."

At the foot of the hill, a yard engine pushed the car up the steep grade onto the ore docks, where each switchman unloaded his car into the freighters far below. Wes smiled again and shook his head. "Railroading was good work in those days. It still is, for those few of us left in it. But back then it was special. It was a great experience to be twenty years old and standing on top of a car carrying ninety tons of ore up onto the docks a hundred feet above the lake. In the evening we could look out and see the ship lights moving along away off on the horizon. Looking out over the lake at those lights, a young fella felt just like the king of the world up there."

On Strike in the Mesabi

The Merritts lost their mines, they lost their railroad, and they died as poor men, while the ore flowed down in an ever-increasing red flood into the expanding furnaces of millionaires.

— Harlan Hatcher, *The Great Lakes*

One day about a century ago, three brothers, timber cruisers with the magnificent names of Leonidas, Cassius, and Napoleon Merritt, were roaming the dense woods of Minnesota's Mesabi territory, one hundred miles northwest of Duluth, when they stumbled on what would turn out to be the single most valuable mineral deposit in the history of this continent. In the swampy uplands near the headwaters of the St. Louis River, just a few feet under their boots, lay a lode of iron ore rich beyond imagining; in places the soft red hematite rocks were actually strewn about in plain sight.

The ore discovered by the Merritt brothers tested an astonishing 64 percent pure iron. In 1890 they purchased a tract of land extending far down into the Mesabi hills, drove first a primitive road and then a narrow-gauge rail line through the woods to Lake Superior, and fell to mining with a vengeance. A million tons of iron ore per week still come out of the Mesabi Range, though the hard-working Merritts eventually lost every penny of their investment to John D. Rockefeller and other eastern magnates. All three of the brothers died paupers and were buried in an obscure corner of a cemetery in Duluth, the city their discovery had transformed from a rough frontier shambles of thrown-together shacks to a major port.

Though less storied, the iron-ore rush to the upper Midwest during the second half of the nineteenth century was greater by far than either the California or the Alaska gold frenzy. Star-struck amateur prospectors raced up to America's North Coast by the tens of thousands, the vast majority of them finding nothing but mosquitoes in the summer and snow, tons of it, in the winter. Among the few who did strike pay dirt was Frank Hibbing. A recent immigrant to the Mesabi area from Germany, Hibbing caught wind one morning of a rumor that a local well digger had come across some promising red soil in a shaft near a lumber-camp bunkhouse. That same afternoon Hibbing was on the site and digging. The red stuff was iron ore, all right, and by his best calculations, the deposit stretched all along the spine of a ridge to the tiny outposts of Virginia City and Chisholm. Following his hunch, he discovered the richest deposit of all, known today as the Hull-Rust Mine — into which I now found myself staring, with utter incredulity.

Here it was, the largest open-pit mine on the planet: a man-made chasm four hundred feet deep, a mile wide, three miles long, and as deserted this afternoon, during the miners' strike, as the Grand Canyon in a prehistoric dawn. The immense excavators, each weighing three hundred and fifty tons and able to chomp out fifteen tons of ore at a single bite, perched far below on the terraced slopes like toy steam shovels. Dump trucks the size of four-story apartment buildings sat silently beside them. The roads twisting down the sides of the pit from terrace to terrace were empty; except for a lone security guard standing nearby on the viewing platform there wasn't a soul in sight.

Just ahead, on a flat stretch of road heavily wooded on both sides, I spotted what looked like a small carnival. Automatically I slowed down. A big yellow, white, and blue tent, open at the sides like a pavilion, was pitched in a field near the junction of the highway and the mine entrance. Cars and pickups were lined up on both shoulders. Under the tent men and women in white aprons were cooking hot dogs and hamburgers over barbecue grills, other men were playing horseshoes and drinking beer out of cans, and somebody had a boom box blaring country music. As I pulled over, two cars passing in the opposite direction blasted their horns, and a few of the men near the road held up cardboard placards announcing that the United Steel Workers local was out on strike.

I walked over to the horseshoe pit, just downwind from a veil of smoke drifting out of a rusty barrel burning damp chunks of spruce to keep off the mosquitoes. Immediately five or six of the striking miners came up, shook hands with me, and introduced themselves. When I inquired exactly what the strike was about, a maintenance mechanic at the Hull-Rust Mine told me in a few words. "Our benefit package," he said. "Our retirement reserve fund gets lower every year."

"How are the negotiations coming?"

"Badly. We've been out several weeks now and haven't made any progress. The problem is, the mine's outside management firm won't come up here and sit down and talk to us face to face. They just keep threatening to light out to Mexico or South America and open new mines there if we don't go back to work on their terms."

More strikers drifted over, seeming almost to materialize out of the acrid smoke from the barrels. The mechanic thrust an ice-cold beer into my hand. "I'm going to put this bluntly for you, mister. The owners and their hired management are involved in a joint venture here to screw the little men, just the way old John D. Rockefeller screwed the Merritt boys back a hundred years ago. But what really gets to us is we don't know the people we're working for anymore and they don't know us. It's a well-known fact in the mining industry that we run the most efficient open-pit operation in the world here. We're proud of it all from A to Z. What we'd love is a chance to show the new back-east management just exactly what we do and how. Take 'em around so they can see for themselves, hear the story of the Hull-Rust Mine from the horse's mouth, so to speak."

"So why won't they meet with you?"

"They're afraid they'd begin to see our point of view," he said. "It's a lot harder to beat a man out of what's rightfully his after you've met with him face to face and heard his views firsthand."

"There was a story told out here in the Mesabi that the four toughest places in all history were Seney, Ely, Hibbing, and Hell, and Hibbing was the worst of the bunch," Herbert Grew was saying to me half an hour later. "I can vouch for it. I was Hibbing's town constable back in the 1950s, when the Hull-Rust Mine was expanding, and let me tell you, every night was Saturday night in those days. It wasn't unusual to see half a dozen different fist fights raging up and down the main street simultaneously."

We were admiring Hibbing's Iron Man, an eighty-one-foot-tall cast-iron sculpture commemorating the Hull-Rust ore discovery nearby, and Herbert Grew, a wiry-looking man in his sixties, was giving me an oral history of the Mesabi. "My grandfather came into this area back before the turn of the century, about the same time as the Merritt brothers. His name was Francis Grouloux, and he was originally from French Canada. After he got to the States he Americanized his name to Grew. My grandfather Grew was a timber cruiser. My father operated a locomotive for the old Oliver Mining Company, and I've done a little bit of everything. Being constable kept me on my toes but I liked it, nothing dull about that job at all. Talk about stories! We had huge gangs of itinerant construction workers coming and going in those days, the jail was packed every night. Hibbing was full of hotels, and we had a saying that the chambermaids were there to use the linen, not change it. There was a real frontier atmosphere."

"Speaking of frontiers," I said, "your country up here, your big Minnesota woods and mineral deposits and lakes and rivers — they're all hundreds of miles north of the forty-fifth parallel. Where I live, that forms the border. How come all this country doesn't belong to Canada?"

"That's simple. England gave it up after the Revolution, at the Paris Peace Conference. The negotiators, including Ben Franklin, used an erroneous map showing Lake of the Woods, up to the northwest of here, draining *into* Lake Superior through the Rainy and Pigeon rivers. So they made those rivers the boundary with Canada. Actually, Lake of the Woods drains *north* into Lake Winnepeg. If the negotiators had known that and pushed the border south just a hundred miles, our Iron Man here would be standing in Ontario right now and so would you and I. But if you're looking for Mesabi history, go see Violet Asuma at the Minnesota Museum of Mining, just across town. She was *born* in an iron mine and knows everything there is to know about this part of the country."

At the tail end of this Friday afternoon before Labor Day weekend, Hibbing's mining museum was nearly empty. Violet Asuma had just given the last tour of the day to half a dozen visitors from Germany, but she seemed happy to stay on a bit longer to talk with me.

"Right there I was born," she said, smiling, and placed her index finger squarely in the middle of an aerial view of the Hull-Rust pit.

"Your mother worked down there? In the mine?"

She smiled again. "When I tell this to people they all guess something different. But no one ever figures it out. You see, there *was* no mine there when I was born. Just the little village of East Hibbing. Then in 1924, one year after I came into this world, they discovered the iron right under our houses and moved the whole town two miles away."

At seventy, Violet Asuma was a slim, energetic, graceful woman, a beguiling talker with a lingering wisp of a Finnish accent. "My father, Samuel Sillanpaa, had an interesting story, too. He came here from Finland in 1918. But not through Ellis Island. To pass through Ellis Island in those days you needed to prove that you had a marketable trade. In Finland, you see, he'd been the town blacksmith. An honorable profession. But not so much need for blacksmiths in America in the Roaring Twenties, eh? So he went to Canada instead, to Halifax, and made his way west on the freight trains. He smuggled himself back into the U.S. by swimming across the Rainy River from Fort Frances to International Falls. Friends brought him here to the Mesabi and got him work as a blacksmith, shoeing the mules in the underground mines.

"In those days most of the iron mining on the range was done underground. In the mine where my father worked they went down thirty-five hundred feet under the surface: twice as deep as the Empire State Building in New York City is tall! They had to pump the fresh air down to the miners and pump the water out of the mine, day and night. But deep in the ground it is always wet and always cold, and my father developed a constant chill. I never remember a time when he wasn't cold. Winter and summer he sat by a red-hot stove all evening, so close you wondered how he could stand it. He said he was soaking up the heat for the next day. But he never seemed able to soak up enough."

I asked Violet Asuma if her mother had been Finnish, too.

"Oh, yes. From Helsinki. Very poor, like my father, but she had the city culture, the music, the nightlife, the dancing. My mother was a wonderful dancer. She met Sam at a dance the year after he came here, 1921, and to catch my father for a husband, my mother put the red iron ore on her cheeks for rouge. And it worked! He fell in love with the wonderful girl dancer from Helsinki with the bright red cheeks, and they were married. Come now. This way. We descend to the mine."

I followed Violet down a flight of steps and through a simulated shaft to a diorama of an underground mine head. In the dim light, a life-size mule was hauling an ore car on narrow rails past a miner, pick raised, with the reddish iron in the wall of the mine illuminated by the rays of the miner's headlamp. It was a grim scene, with no touch of the picturesque about it.

"Now I will tell you what I don't so often tell," Violet said. "Whether you put it into the book you are writing is up to you. Far down beneath the surface of the ground, far from the sunshine and the fresh air, for twelve hours each day, six days each week, my father carried a seven-pound battery on his hip. This battery powered the carbide lamp on his helmet. In time he became lame from the weight, and he had to give up the dancing. But my mother loved to dance so much that he still took her to all the dances, and while she danced, he watched my twin sister and myself."

In the dim light from the diorama, I was scribbling fast in my pocket notebook. "You think you may put this in your book?" Violet said.

"Yes," I said. "Definitely."

"Then I will tell you the rest. My father was very easygoing by nature, yes? He wanted my mother to be happy. But my twin sister and I, we cried to see her dancing with the other men. Then our father gave us candy to quiet us, so we quickly learned how to cry for our candy. But we knew something else, too. The way twins do, without needing ever to discuss it? Even when we were very small we somehow knew that our mother, being from the city, believed herself to be superior to our father, the Mesabi miner. That, I think, is what truly made us cry, and the true story of Sam the blacksmith, who swam the Rainy River into America and married the city girl with red ore on her cheeks!"

Notes from the Land of Ten Thousand Lakes

There's something exhilarating about dwelling in a place where people take their identity more from where they live than from what they do.

— Chris Braithwaite

4:00 P.M. Since leaving home, I've been on the lookout for indicators of what I think of as true North Country. Good brook trout fishing is one such indicator. So is a notoriously severe climate. Yet another is static over the car radio. Whenever I can't pick up anything but static, I know I'm really up in the hinterlands, exactly where I want to be. The first place this happened was late the following afternoon, just south of Ely (pronounced *E-lee*, long e, accent on the first syllable). Approaching this town on the south edge of Minnesota's Boundary Waters Canoe Area, I can't bring in a single station on either the AM or the FM band of my radio. Even in the Madawaska Republic of Maine and up on Michigan's primitive road, I was able to find something to listen to. But north of the Mesabi, the airwaves carry nothing but crackling static.

5:00 P.M. Here in the wooded hills overlooking Vermilion Lake, the season is both behind and ahead of the rest of the North Country. Fireweed is still in full bloom in roadside clearings; but the hillside maples are already starting to color up, and it's noticeably chillier. In a week or two, leaf-peeping visitors will be flocking here from all over Minnesota.

In 1866 the allure of Vermilion Lake was gold, tons of it, lying right on the surface like the red ore of the Mesabi, easy pickings for thousands of footloose Civil War veterans, hard-bitten Yankees fleeing played-out hill farms, and city boys just off the streets of New York and Chicago and Philadelphia, who charged up to the woods of northern Minnesota only to discover that all those glittering yellow rocks were iron pyrite: fool's gold.

At the feverish height of the Vermilion Range gold rush, an enterprising woman named Daisy Redfield opened a high-class whorehouse in Ely. Soon its fame had spread all over the upper Midwest. Rumor had it that a resourceful young woman with a strong work ethic and an aptitude for rudimentary acrobatics could make a small fortune there in a few short months. By the same token, a prospector could lose his whole boodle on Daisy's premises in a single night. Before long the up-and-coming young Chicago evangelist Billy Sunday caught wind of the shenanigans at Ely; and in the grand old tradition of evangelists everywhere, who seem attracted to fornication the way Daisy's clients were attracted to all that iron pyrite, the Reverend Mr. Sunday hopped a train north. In a marathon of fire-and-brimstone invective perhaps unsurpassed in the annals of frontier ministries, he inveighed against Daisy night and day for a solid week — after which she announced that the famous preacher's free advertising for her pleasure palace had helped business more than "a free barrel of sloe gin."

6:00 P.M. Ely. Where Billy Sunday once preached hellfire on the muddy street in front of Miss Daisy's, cars toting forest-green canoes on their roofs now line the curbs. Yet even today, crowded with people who look like walking advertisements for L. L. Bean, the main street of Ely retains its frontier look. It's lined with outfitters' shops, whose names emblazoned over their show windows have the stirring ring of the North: Wilderness Outfitters, Canadian Waters Outfitters, Basswood Trading Company, Voyageurs North, Northwind, North Country Canoe Outfitters, Canoe Country. These days there seem to be more outfitters in Ely than grocery stores and gas stations. You can pick up a sixteen-foot Old Town canoe or an assortment of multicolored trout flies in this town at any one of half a dozen establishments. If you feel like spending an afternoon strolling through a mall, on the other hand, you'll have a long way to go.

* * *

7:00 P.M. A veteran Ely outdoorsman, the great-grandson of a local Chippewa chief, to me over supper that night: "There's still a lot of woods out there in the Boundary Waters, and they haven't changed greatly from the time my great-grandfather lived in them. But the National Forest Service overregulates the logging, so we're getting more older-growth evergreen forest that's not right for the deer. What's more, the Forest Service wants to reduce the size and number of parties that can use the wilderness and do away with motorized portages, where three-wheelers tow canoes from lake to lake. I have a handicapped friend, a local man, who depends on motorized portages. So do many elderly folks. A growing number of people in this area think there's a conspiracy of outside, arm-chair environmentalists to cut local folks off from the land. Those of us who live here year-round feel disenfranchised. We're beginning to feel that some of the preservationists from outside the area want to save the wilderness, all right. For nobody!"

7:30 P.M. A Duluth businessman, whose four-wheel-drive Scout displays both Sierra Club and Izaak Walton League decals on the rear bumper, to me, in our motel parking lot at dusk: "To tell you the truth, I'm fairly conservative in most of my political convictions. But tomorrow at dawn I plan to canoe out into the Boundary Waters. I'll camp by a waterfall I first laid eyes on forty years ago, and in all that time it hasn't changed one bit. You just can't say that about very many places today. If the preservationists can save that waterfall for another forty years, God bless them. I'll gladly pay my dues."

Canoeing the Boundary Waters

At last I caught what I was listening for — the long-drawn quavering howl from over the hills, a sound as wild and indigenous to the north as the muskegs or the northern lights. That was wilderness music.

— Sigurd Olson, *Songs of the North*

At five-thirty the next morning I was paddling a canoe into the Boundary Waters north of Ely, following the same route that the Minnesota nature essayist and preservationist Sigurd Olson had taken hundreds of times. Before I'd gone a mile, an eagle came soaring out of the north, and with him a fresh dawn breeze and sunrise. The mist tattered apart. Patches of blue sky appeared. And the eagle, his snowy head gleaming in the sunshine, wheeled one hundred feet above the water, looking for a fish breakfast.

By the time I entered the lake's outlet there wasn't a cloud to be seen. The river water was amber-colored, stained by tannins leached out of the evergreen forest. It looked trouty, but I was searching for something else: a wolf or a deer kill left by a wolf or even a wolf track. A nearby wooded ridge looked inviting. From its crest a big gray (as timber wolves are called in northern Minnesota — their last wild bastion in the lower forty-eight states) could bask in the sun and keep tabs on a vast swath of territory. I had a compass and a pair of waterproofed boots. Tramping up there to see what I could see was worth a shot. Why not?

I nosed the canoe into a sedgy marsh, took a quick bearing, and struck off dead west toward the ridge. For the first few hundred yards the footing was spongy. A beaver run branched off to my right, and I thought of the Maine game warden Fred Jackson, up to his nose in just such a backwater, waiting for oncoming Acadian poachers. I checked the muddy bank of the run for wolf tracks, found only the big webbed prints of a beaver, and continued toward the ridge.

Soon I entered a band of spruce and fir trees draped with the gray-green hanging moss that deer love above all other food. Higher up the ridge, in a mixed woods of evergreens, maples, beech, and birch, I spotted the unmistakable two-pronged indentations of a small deer, probably a yearling, headed down toward the river. Nearby I counted several varieties of ground cover: white caribou moss, springy green ground pine, and an ankle-deep, resilient, hemlock-colored moss I couldn't identify. Now I was scrambling around granite outcroppings, some pink, some silvery dark. The woods glistened from an overnight shower; even the spider webs sparkled brightly.

Getting out into the northern wilderness never fails to produce in me that exhilarating sense of well-being that I first experienced on my early fishing excursions with my father and uncle. And here in the tall Minnesota woods, I was excited to discover rose-colored bunchberries, the glossy blue-black berries of clintonia, fiery red partridgeberries, and, in a ridge-top clearing created by a forest fire years ago, blueberries as big around as the end of my thumb. The view in all four directions was a green-and-blue amalgam of rivers, lakes, ponds, islands, peninsulas, and dark wooded isthmuses. A mile to the north I could see the Pigeon River, which forms the Canadian border in this part of Minnesota; but no trace of wolves yet.

The sign was set back at the edge of the woods near the junction of the Echo Trail, a wilderness road on the southern fringe of the Boundary Waters, and a steep, red-dirt road forking off to the north. It was small and hand-lettered, but in the bright late-afternoon sunshine it immediately caught my attention: TIMBER WOLF POINT.

Although I'd continued to look hard all day, even going so far as to stop my car where streams crossed under the Echo Trail to check the sandy shorelines for tracks, I'd seen no indication that wolves lived within a hundred miles of here. Timber Wolf Point might be as close as I would get to a real wolf. The dirt road looked passable, though just barely, so I went for it, figuring that if the track got too bad I could always park the car and walk.

In fact, the narrow bush road was just half a mile long, ending on the east side of Echo Lake at Timber Wolf Point Resort, no farther than a long rifle shot from the international border. Larry Oberg, who ran the resort with his wife, Kathi, met me at the door of his log office, and although they were out straight, changing linen and cleaning guest cabins for the upcoming Labor Day weekend, they dropped everything and invited me in for coffee. When I told the Obergs that I'd been looking for a wolf, Larry, who looked much younger than his late thirties, said that his father had run a trapline in the Echo Lake area back when it was all uninhabited wilderness. Within sight of where we were sipping our coffee, he'd caught a pure black timber wolf, eight feet long from the tip of its jaw to the tip of its tail. Larry's dad named the spot Timber Wolf Point, built a trapper's cabin there the following year, and in the early 1960s established a small hunting and fishing lodge on the lakeshore.

"We had no running water, no electricity, no indoor toilet," Larry said. "We cut our own firewood, and to keep warm we had to bank our cabin with snow up to the windowsills. It was terribly primitive, but we had the dream of the resort, and we worked hard at it day and night for years. The road you came in on? It didn't exist in those days. During the school year, my brother and I took our boat across the lake and walked through the woods to the nearest highway to meet the school bus. In the winter we got across the ice on an old snowmobile. Mom had just two rules for us. She didn't let us go to school when the temperature dropped under forty below or during a whiteout. An ordinary snowstorm was okay, we could handle that all right. But you can get lost and freeze to death in an hour in a northern Minnesota whiteout."

Just how cold could the winters here get? Larry laughed. "Our thermometer goes to sixty below. More than once, we've seen all the mercury concentrated in a tight ball below that."

Kathi Oberg, a pretty, dark-haired woman who looked even younger than Larry, said that she'd been brought up down in Minneapolis. "I was a typical city kid who didn't know anything at all about country life, much less North Country life. The first few winters after Larry and I were married were so long and tough, I almost didn't adjust. I could really identify with the people in Laura Ingalls Wilder's Little House books, especially *The Long Winter*. Now I wouldn't live anyplace else. The North Country gets in your blood, you know. Sometimes it's all you can do to manage to get by in it, but after a while you can't live without it. I'd even miss the wolves!"

Larry and Kathi saw wolves frequently, especially in the winter, when the big grays sometimes ran down deer in front of the lodge on the frozen lake. Larry said, "We admire the wolves, and yes, we'd miss them, but it's hard to feel the same way about them that you do about your pet dog. They aren't dogs, and though they can be tamed to a degree, they really aren't pets. Wolves don't kill for sport, but we've watched them hamstring deer, drag them down, eat out their heart and liver, and leave the rest. As I told Kathi when she first came up here, it's not a pretty thing to see, but it's the law of the North."

We walked out to the end of the point and looked across Echo Lake toward Canada. From this perspective, the lake seemed as wild as it was when Larry Oberg's father set up his first trapline here. "This is our dream come true," Larry said. "Dad's, Mom's, mine, and now Kathi's. But to tell you the truth, we don't know what the future holds for resort owners like us. Trouble is, resorting is an unpredictable business. We don't get any subsidies or government help. Just higher taxes, and more regulations all the time from federal agencies that can't even agree with each other. When Dad started out up here, there were about five thousand fishing resorts in Minnesota. Today there aren't a thousand."

Larry picked up a flat stone about the size of his palm and studied it. "Don't mistake me. I'm not pessimistic about keeping going, and we aren't going to throw in the towel. On the bottom line, what we're doing up here isn't just about a business. It's about a way of life well worth holding on to."

By now it was late afternoon, and I was eager to get on to International Falls. I had only a few more miles to go on the Echo Trail before leaving the Boundary Waters, and I was moving along at a fair clip, not hurrying but driving purposefully toward my destination when ahead, on the right shoulder of the woods road, the wolf simply seemed to materialize, long-legged and dark as night. It appeared so abruptly that I actually didn't have time to be surprised. Like the eight-foot monster Larry Oberg's father had trapped just a few miles away, it was pure black, and it moved with a feral, sidling lope, completely unlike a dog. As I skidded to a stop, it trotted into the woods. But instead of vanishing into the forest, it suddenly veered and ran back toward my car. No more than twenty feet away, it paused and looked straight at me for perhaps ten seconds.

No camera could have captured that wolf's wild expression, much less my own sense of excitement at this face-to-face encounter with the animal which, more than any other, was emblematic to me of the Boundary Waters. There it was, and there I was, entirely by chance. I'd gladly have driven all the way from Vermont to northern Minnesota for this experience, which meant everything to me and nothing at all to the wolf. Now, as suddenly as it had appeared, it faded away into the trees, leaving me alone on the woods road with as fine a memory of my day in the northern Minnesota wilderness as I could wish to possess.

The Northeast Kingdom

This is a beautiful part of Vermont. It should have a special name — the Northeast Kingdom.

— Senator George Aiken, 1952

Driving west out of the Boundary Waters territory after my encounter with the timber wolf, I knew exactly how Larry and Kathi Oberg felt about staying on in their special stretch of the frontier. I'd felt exactly the same way about Vermont's Northeast Kingdom, since the warm evening in late May of 1964 when I first laid eyes on it.

It was dusk, the day after Phillis and I graduated from Syracuse University, and we were standing on the bank above the falls on northern Vermont's Willoughby River, a few miles south of the Canadian border, where we'd come to interview for teaching jobs at the small local high school. Suddenly, in the luminous spring twilight, a run of Canadian rainbow trout, their crimson sides flashing, came rushing up through the cataract below us, leaping high out of the water to clear the lip of the falls on their way upstream to spawn. After driving through mile upon mile of woods to reach this tiny border village, I had thought I was prepared for nearly anything; but I could scarcely believe my eyes at the sight of these huge, leaping trout.

"Do you like to fish?" the superintendent of schools asked me.

"Only for trout, and only every day," I said.

He nodded. "All I can offer you is forty-two hundred dollars apiece," he said, half apologetically.

I looked at Phillis, who smiled. "Why not?" she said, and that was that — in a matter of sixty seconds, we'd found our lifetime home.

When we came to the Northeast Kingdom in 1964, both Phillis and I knew that we wanted to live in the country, preferably near the border. Now, nearing the end of the second leg of my cross-country journey, in a similarly remote territory, having just visited a couple who for many years had made their living operating a fishing resort, it occurred to me again how fishing, more than any other activity, had connected me with that harsh and lovely remnant of an earlier Vermont.

"Trout live in beautiful places," Robert Traver wrote in *Trout Madness*, his anecdotal celebration of a lifetime of fishing in Michigan's Upper Peninsula. "And that is why I love to fish for them." Vermont's Northeast Kingdom, when Phillis and I moved there, was full of trout and the beautiful places where they live. One such place was the vast cedar bog that our landlady, Verna Fletcher, directed me to one Saturday morning during our first fall in the Kingdom.

"The woods are so thick in there you'll hear the brook before you see it," Verna said. "When you get to the fork in the deer trail, start listening."

A couple of hours later, Phillis and I were creeping through big spruce and fir woods, listening hard. Finally we heard the lightest rippling noise. The wind? No, it was definitely the brook, spilling over a beaver dam on the south edge of a flooded bog that seemed to stretch all the way north to Canada. This lovely and spectral place, which I would return to hundreds of times over the years without ever once seeing the track of another human being, would later become the setting for several of my first stories. At the time, it was merely a wonderful place to fish, teeming with large, splendidly colored brook trout.

Later that fall I listened to Verna tell how, during the Depression, she'd saved her family farm on the ridge overlooking that bog by working in the barn and fields fourteen and sixteen hours a day and then, under the cover of darkness, slipping down to the brook and manufacturing white lightning, which she sold in the village where we were currently teaching. She told us that late one night she looked up from her whiskey still to see by the light of her lantern a stranger in a suit and tie — a revenuer! When she told him that she was using the income from the moonshine to pay the mortgage on the farm, he simply faded back into the night woods without speaking a word. Years later, after her first husband died, he reappeared at her door. She knew him instantly. "Have you come again to arrest me?" she said.

"No, I've come to marry you," he told her seriously, though two years after they were married he, too, died suddenly.

"I ain't had good luck with husbands," Verna told Phillis and me matter-of-factly. "You two get along good because you do things together. You fish and hunt and go to fairs together and enjoy one another's company. Phillis, you keep going fishing with Howard. Howard, you take Phillis to Montreal and Boston and the ocean when you can. Keep doing things together, and you two will make out fine."

It was golden advice from a wise North Country woman, which I still think of when Phillis and I strike out together for a Northeast Kingdom trout brook or a weekend in Montreal.

I didn't get around to writing Verna's own story until a few years later, after attempting a road novel based loosely on a cross-country hitchhiking trip to the Pacific back when I was in college. From that project, which I hammered out nights and weekends and summers during my early years of teaching in the Kingdom, I learned a thing or two about how to write an extended narrative. But from Verna and our other Kingdom neighbors, I would learn what I should be writing fiction about.

Fishing continued to play a large part in my acclimation to the Kingdom. As soon as my students found out that I liked to fish, I had more invitations to go fishing with them and their families than I could accept. Many of these excursions resulted in good story material. Soon after school started, as a result of a class discussion of Hamlet's speech beginning "There are more things in heaven and earth, Horatio . . . ," a quiet, hard-working farm kid took me to a hidden pool at the foot of a mountain falls where a year earlier, he told me seriously, he'd seen the ghost of a drowned cousin. I wasn't sure what to make of his experience, but eventually his story would work its way into one of mine.

Soon afterward I met Margery Moore. I'd stopped and asked

permission to fish the stretch of river running through Margery's hill farm — named, not without a touch of irony, Peaceful Valley — and from that day onward she was one of our closest Kingdom friends. More than half Indian, with striking blue eyes, Margery had traveled extensively in her youth, working as a lumber camp cook, attendant at a mental institution, housekeeper at a millionaire's estate, rancher, barkeep, and social worker. When the local bank was robbed in the late 1970s and the FBI, for lack of any real leads, homed in on Margery's Peaceful Valley and her tenants, she tricked them into moving her twenty-year-old manure pile on the assumption that the forty thousand dollars from the bank heist was buried at its bottom. Next she had her grown sons enter a car at the local demolition derby emblazoned with slogans thanking the local bank for its generosity; as the old clunker roared onto the track, her boys hurled thousands of dollars of Monopoly money out of the windows, to the delight of half the county. Over time Margery told me her own life story and scores of others, many of which eventually made their way into my fiction.

I never could persuade my longtime literary mentor, the lyric poet Jim Hayford, to go fishing with me; but from the start, Jim encouraged me to write about my fishing expeditions and the students and old-timers I went with, to steer clear of academic life, and to stay in the Kingdom, as he had. A Vermont native, Amherst graduate, and student of Robert Frost, Jim was a gifted teacher who had been blackballed for his support of the socialist Henry Wallace during the 1948 presidential election. In the Kingdom he had raised goats, worked as a carpenter and itinerant piano teacher, edited grammar texts, driven a taxi, and written and published more than six hundred first-rate lyric poems in the tradition of Frost, Dickinson, and Emerson.

Like everyplace else, the Kingdom had its dark side. Larry Curtis, who investigated the arms smuggler Gerald Bull, could tell you about that. So could the black minister, recently arrived from California, who in the summer of 1968 was driven out of Irasburg, the village where Phillis and I first went fishing in the dawn after our wedding day, and where we now make our home. After an unprovoked shotgun attack by drunken nightriders on the minister's house, the local police, abetted by a racist newspaper editor, investigated the victims of the shooting rather than the nightriders. The minister was subsequently hauled into court for "committing adultery with a white woman" and harried back to California, while the local shooters were charged with nothing more serious than disturbing the peace. Another fifteen years would pass before I wrote about this tragedy, to which I added a murder and some extensive courtroom scenes. The real key to my novel *A Stranger in the Kingdom*, however, was the trout-fishing Kinneson family, which in many ways resembles my own extended family from my boyhood in the Catskills and on my grandparents' farm.

Like Larry and Kathi Obergs' frontier, the Kingdom was a special and unique place, a place apart from the rest of Vermont and New England, full of fiercely antiauthoritarian, independent-minded individualists, like Verna Fletcher and Margery Moore and Jim Hayford, people with original stories to be told. I fell in love with it the evening we first stood on the bank of the Willoughby River and watched the big jumping trout.

Yet to understand the Kingdom well enough to write about it, I knew that I would have to learn more about its darker side as well, and come to terms with it in a way that I could not do as a high school teacher.

Part Three

BORDER TOWNS

Food Gas Grocs Beer, Minnesota

You don't see many homeless up here in these parts. Poor, we're all poor. Drunks, unfortunately we've got our share of them, too. But homeless you aren't very apt to run into in the North Country.

— Pool player in a crossroads store
somewhere in northern Minnesota

FOOD GAS GROCS BEER declared the handwritten sign nailed to the outside of the dilapidated building at the end of the Echo Trail. Half a dozen cars and pickups were parked in front of the place, and a single gas pump leaned off at a precarious angle in the puckerbrush beside it.

"Keep your gas tank topped off every chance you get" was probably the best practical advice I received from Maine to the Pacific, so although by North Country standards I was just a hop, skip, and a jump — maybe seventy miles — from International Falls, and I still had a third of a tank, I pulled in and filled up.

The store consisted of a single low-ceiled dusky room, rank with the stale odors of cigarette smoke and beer. The "Grocs" consisted of a couple of shelves of bread and canned goods and a squat old refrigerator, with a motor on top, containing a dozen or so six-packs of beer. A couple of young guys in baseball hats were shooting pool at a table in the middle of the room, and a very young woman with a sleeping baby flopped over her shoulder was watching them. Along the back wall ran a makeshift bar where three men hunkered on stools drinking beer out of long-neck bottles and watching a big television set. I paid for the gas, sat down at the bar, and ordered a beer.

TV reception was lousy up here, but the men were intently watching a fuzzy show on hunting in the Yukon. A red-faced man in his sixties on the stool beside me tapped my elbow with the back of his hand and pointed at the blurry screen. "Big country," he said.

"It is," I said. "You've got some pretty big country right here in Minnesota."

He cranked himself around on his stool and studied me for a moment, not impolitely. I was wearing jeans, work boots, a wool sweater over a flannel shirt.

"Been out in the bush?" the red-faced man said.

"Not very far. Just up to the border and back."

He nodded again, and in the tone of a man launching into a much-practiced set piece, he announced, "Well, one thing's for sure. You can't get lost in the bush. Not if you know what you're doing out there."

The two men beside him were grinning.

"Oh, no, you cannot," he said in a more strident voice, looking around to see who was listening. "If you know what you're doing and keep your head, you cannot get lost in the bush."

He glanced back at me. "Do you think you can?"

"I might manage to," I said. "I'm from Vermont. I don't know this country."

"Where you're from don't matter. If you have a compass, you can't get lost. Not here, or" — he gestured at the television screen — "there."

"Harley," one of the pool players said, "have you ever been to the Yukon?"

Harley set his bottle down on the bar. "You can't starve in the bush, either," he said quite defiantly. "Not if you have a rifle and matches."

"I'd hate to be out in the Yukon when the temperature's seventy below zero, I don't care how many matches I had," the man beside him said. "All that would keep me from starving is I'd freeze to death first."

"You would not," Harley said. "A man with matches and a rifle and a compass can always survive in the bush."

As I reached for my wallet, Harley began to sway on his stool. "I never been los' in my life," he said, and put his head down on the bar and passed out.

One of the pool players shook his head. "Good old Harley," he said. "He's a piece of work."

"Was he a good woodsman in his day?" I asked.

The pool player set down his cue and picked up a long-neck on the corner of the table and took a swig. "No, sir. So far as I know, and I've lived here all my life, Harley's never been nothing more nor less than the town drunk. We rag on him some to keep him in line, but we put up with him. Make sure he don't run out of firewood in the winter, has something to eat in the house, a place to go on Thanksgiving and Christmas. What it is, see, this is his home. Same as it is ours."

"You've got a town drunk," I said. "Where's your town?"

The pool player laughed. "Right here. Place is owned by a fella named Smitty, but we call it 'Food Gas Grocs Beer, Minnesota.' It's all that's left of the town that used to be here. It and Harley, the town drunk."

As I forged on toward International Falls through the big woods north of Food Gas Grocs Beer, I thought about the tiny border towns I'd traveled through from Lubec, the dying fishing village on the coast of Maine, to Food Gas etc., Minnesota, which had, it seemed, died already, like my hometown of Chichester. An observation from an essay by Wendell Berry came to mind. In "Word and Flesh," Berry writes that his "small community in Kentucky has lived and dwindled for at least a century under the influence of four kinds of organizations: governments, corporations, schools, and churches — all of which are distant (either actually or in interest), centralized, and consequently abstract in their concerns."

What struck me as remarkable in this respect about Food Gas Grocs Beer was that although the place had long ago lost its own government, businesses, schools, and churches — lost its whole architectural identity as a town, actually — it had somehow managed to retain some of its deeper character as a community. But how? On the spot, I decided to spend some of the following days looking into the current state of border towns and communities, finding out what defined them as either or both and made them good or unsatisfactory places to live in.

The Coldest Spot in the Lower Forty-Eight

No border town is anything but a border town.

— Raymond Chandler, *The Long Goodbye*

I grew up in International Falls, and when I was a little girl, Native Americans used to come over the bridge from Canada with blueberries and fresh fish to sell. Today many of the Chippewa women, including a good number of older women with grown-up kids, cross the bridge every day to take classes at the Rainy River Community College. Now *that's* what an international community should be all about, in my opinion.

— Barbara Wood, U.S. immigration supervisor

6:00 P.M. International Falls, Minnesota. This town has a raw look that I like at first sight. The Rainy River, marking the border, looks cold and surprisingly wild, not much different from the way it must have looked when Sam Sillanpaa, the Finnish blacksmith, swam across it to the United States decades ago. Just before setting out on a walking tour, I fortify myself with a cup of coffee and a hamburger at a main-street café with a stamped tin ceiling and big old-fashioned fans with blades like propellers, air conditioning being superfluous in a town often referred to on national weather broadcasts as the coldest in the lower forty-eight states. On the wall above my table there's a black-and-white photograph of International Falls in the 1930s. From where I'm sitting I can look out the window at the same scene, which has changed scarcely at all in the past sixty years. Back outside, I have the eerie sense of stepping into the much earlier town in the photo.

6:30 P.M. Down at the west end of the street, in a small park across from the high school, stands a thermometer as tall as a house painter's extension ladder. It reads fifty-three degrees, though with a sharp wind gusting across the river out of Canada, the chill factor must be down in the thirties. Very possibly, tonight will be another record low.

6:45 P.M. There's a commotion taking place at the American end of the bridge connecting International Falls with Fort Frances, its Canadian sister town across the river. An eighteen-wheeler carrying wood chips has somehow jackknifed between the American customs station and the median. Officials are rushing about; stalled motorists are blasting their horns; and Canadians returning home are laughing and calling out facetious advice to the truck driver in French and English. In the midst of the hubbub, one slender middle-aged woman wearing a U.S. Immigration Service uniform and a wryly bemused expression is patiently directing the driver in his attempts to extricate his rig. He has no more than three inches of maneuvering leeway on either side, and the officer has to keep reassuring him not to panic, that yes, he'll eventually free himself — as he eventually does, driving off to a chorus of mock cheers and horn honking from the gallery of motorists.

7:00 P.M. So I introduce myself to the official who helped the truck driver: Barbara Wood. When I congratulate her on her equanimity during the crisis, she laughs. "That's what I like about this job. A new challenge every day. Back in July we had ten thousand members of the Chippewa Nation out on the bridge celebrating Jay's Treaty. It all went fine, a great success really, though speaking just for myself, I think that some Native Americans are misinterpreting one of the provisions of the treaty. It does grant Indians free access to cross into and out of the U.S. anytime they want, but doesn't exempt them from paying duty on goods brought across the border. Still, it's heartening to me to think that all of us up here, Americans and Canadians and Native Americans on both sides of the border, live in a place where such historical considerations are still an important part of our lives. So many people these days are totally cut off from the history of the places where they live."

* * *

7:30 P.M. Border Bob's Emporium, just off the international bridge. Border Bob's looks as if it was once a warehouse for storing lumber, maybe, or possibly a feed store. These days it's a labyrinth of bright wool shirts and pants, fishing tackle, hunting rifles and shotguns, fiberglass compound bows and hunting arrows, hiking boots, insulated winter clothing, hardware supplies, maps, canoes, paddles, tents, and whatever else you'd need to dwell outdoors in the North Country for as long as you wanted to. Then just when I think I've finally seen everything, I emerge into a spacious workroom resembling nothing so much as a natural history museum filled with the most lifelike mounted fish and game animals I've ever seen.

Regarding me from the walls are the heads of moose, caribou, and deer, all sporting trophy-sized racks. In glass-fronted wall cases around the room are minks and otters and beavers mounted in striking poses. There's a Canada lynx with great snowshoe paws and fierce-looking ear tufts three inches long. There are timber wolves, and ducks and geese, and fish by the dozen, pike and bass and salmon and gigantic muskellunge, glistening as though they'd come out of the water seconds ago. And at work in the middle of the room, near a bench covered with tools of all sizes, from broad-headed hatchets and hacksaws down to tiny lances and scissors as odd-shaped and esoteric as a surgeon's instruments, a tall dark-haired man in his thirties, sculpting a standing bear, front legs extended, from a styrofoam block about eight feet high: Border Bob.

8:00 P.M. Border Bob Neuenschwander: "I started out as a taxidermist in 1982. It's amazing how far the technology of the craft has come in just that time. These days almost everything is done from a model, with the skin stretched over it afterward. I tan a bear's skin and head, take the measurements of every last detail, and sculpt a model to exact scale for the skin to go over. The term 'stuffed' doesn't really apply to most taxidermy mounts anymore. I don't even stuff fish. I make a graphite model and stretch the fish's skin around it. For catch-and-release fishermen, I build the model from a photograph of the fish and paint it. I just wish we could do the same thing with this guy." Bob nods toward the bear model. "Or this one" — pointing at a pure black wolf, posed in midstride, its loping motion caught as perfectly as the brook trout's fiery red belly.

I mention the eight-foot-long black wolf that Larry Oberg's father trapped on Echo Lake and the big dark wolf I saw that afternoon, and ask if black wolves are a separate subspecies in these parts. Bob says no, just a melanistic variety of the local gray timber wolf, and not all that uncommon. His personal theory about the wolves' presence in northern Minnesota and not, say, in equally wild areas of northern Maine is that the big grays simply drifted south with the Polar Express that brings Minnesota its fiercely cold weather. "We actually have about the same type of climate and topography here as the country five hundred miles north of us in Ontario," he says. "The wolves are part of the package."

"So these are by ancestry Canadian wolves?"

Border Bob laughs. "Let's say they're North American wolves and as such part of the international community."

Notes from Warroad

I intend to keep farming. I don't care if I have to take ten part-time jobs to do it.

— Dan Heppner, in Warroad, Minnesota

6:00 A.M. Warroad, on the southwestern end of Lake of the Woods at dawn. The sky seems bigger here. I imagine that I can sense the looming presence, just to the west, of the Great Plains. Warroad itself looks like a ghost town this Sunday morning, so I drive out to the park and public boat launch and go for a tramp along the lake. Although the weather is supposed to turn mild and sunny later in the morning, a sharp wind is blowing in off the water. Yet even on an overcast day, Lake of the Woods looks as wildly beautiful as its name.

7:00 A.M. On my way back to the car, I meet another early-morning walker, a well-built man of about forty in a suit and tie. It's Dan Heppner, a local rancher and farmer, out taking the air before church. I tell Dan that I'm headed up to the Northwest Angle, which I understand to have been homesteaded within the memory of some of its older residents. He laughs. "You heard right about that. One of the things I do to keep my farm going, I'm a substitute school-bus driver. The first time I drove the bus up to the Angle, the regular driver told me to go clear to the end of the road, right up to the water, blink my headlights, and wait. It was October and still quite dark when I got there. I flashed my lights and then I sat. About ten minutes later a canoe appeared out of the mist. I mean it just materialized. Paddling it were two kids from the Red Lake Indian Reservation. The older one wasn't more than twelve, and his sister couldn't have been ten. A minute later another canoe with two kids appeared. Then another. Two little kids in each one, paddling to meet the school bus. I felt like I'd just stepped into the last century."

* * *

7:15 A.M. Abruptly Dan stops in his tracks and turns to face me. His reminiscing smile fades. "I'll tell you something," he says. "I shouldn't be driving a bus at all. I've got four hundred acres of very rich farmland on the edge of town here. But the cost of everything a farmer needs to stay in business keeps rising. Tractor fuel, truck fuel, farm machinery, electricity, you name it. In the meantime, my income from the farm stays about the same. So when the school-bus driving deal came along, I took it. I had to. But I kept looking for a better job to pay our bills, and finally my wife said, 'The trouble is, Dan, that in today's technological workplace you're unskilled.' Well, sir, I took exception to that. 'I'm highly skilled,' I told her. 'I'm skilled at running a four-hundred-acre farm. I can coax wheat and hay out of the ground in a place where the temperature drops to fifty below in the winter, keep twenty-year-old tractors and bailers and combines in good running order on homemade parts, and maintain a sophisticated bookkeeping system. I can doctor a herd of cattle, keep 'em as healthy as any cattle in the state, I read all the latest literature in my field that I can get my hands on, I work from four in the morning until long after dark, and yes, I'm forty years old and not as tough as I once was, and to stay afloat I'm driving a school bus. But unskilled? Unskilled I am not. What's more, I intend to keep farming. I don't care if I have to take ten part-time jobs."

The Northwest Angle

On a map of Minnesota, the Northwest Angle juts like a thumb into the smooth Canadian underbelly at the 49th parallel. A geographical orphan, stranded by a mapmaker's error, the Angle represents the northernmost point in the lower 48 states, a remote spit of woods and water surrounded on three sides by Canada. To the west is Manitoba; to the north and east lie the great dense forests of Ontario; to the south is the U.S. mainland. This is wilderness. Forty miles wide, seventy miles north to south. Gorgeous country, yes, but full of ghosts.

— Tim O'Brien, *In the Lake of the Woods*

To reach the Northwest Angle by road from Minnesota, you have to drive through forty miles of Canada. With a surge of excitement, I crossed into Manitoba and followed the provincial highway north to a rough gravel road veering off toward Angle Inlet. Then I drove mile after mile through dense woods until at last I arrived at a monument just off the road to my left marking the U.S. border and the official boundary of the Northwest Angle. Here the American–Canadian line runs due north and south instead of east and west; there's no customs station, just the lone granite marker and the twenty-foot-wide clearing known all along the border from coast to coast as the Vista. As I got back into the car, two deer with deep red summer coats stepped out of the encroaching woods and began feeding on the wild grasses. A mile farther east, a deer already wearing its charcoal-gray winter coat leaped across the road in front of the car in one bound. Tim O'Brien got it exactly right in his haunting and unforgettable novel *In the Lake of the Woods*. "This is wilderness."

The gravel road branched several times, with no signs to indicate which fork to take to Angle Inlet. Twice, I chose the wrong turn and wound up dead-ended in a maze of inlets, bays, coves, and marshy streams so closely resembling tidal estuaries that I half expected to catch a whiff of salt on the breeze. A canoe would have done me more good than my car here, I thought. This watery place was so totally undeveloped that it didn't feel like either Canada or the United States, but an unspoiled no-man's-land in between. Still, it wouldn't do to romanticize the Northwest Angle, which not so very long ago, once winter set in, was as hard a place to survive in as any in America. In *The Falcon*, a wonderful memoir of thirty years (1790–1820) spent with the Ojibwa of the upper Midwest, John Tanner writes of the territory near Lake of the Woods: "Their barren and inhospitable country affords them so scantily the means of subsistence, that it is only with the utmost exertion and activity that life can be sustained, and it not unfrequently happens that the strongest men, and the best hunters, perish of absolute hunger."

In the course of my musings, I blundered my way into Angle Inlet, which consists of a single café, a few scattered log cabins and houses, and a tiny building with a sign declaring it to be the northernmost post office in the contiguous United States. Not a soul was in sight, so I walked over to the café, where a waitress drinking coffee at the counter advised me to visit Joe Risser.

"If you want to know about the Angle, Joe's the man to talk to," she said. "His folks settled this area."

Over the Angle this morning hung a sabbath quietude, a hiatus in the week that I have never much cared for. Joe Risser, just driving back into his yard from an early-morning visit to a nephew, seemed to be exactly of my mind on the subject. Trim and fit at eighty-two, with large, powerful hands that had obviously labored hard for well over half a century, Joe was a man on the move. He was headed for his woodworking shop; but when I told him I was interested in the Angle, he invited me into his living room, where he'd kindled a small wood fire in his parlor stove to take the chill off the morning.

"One yarn goes something like this," Joe said, leaning over to plink a jet of tobacco juice into a can beside the stove. "The original surveyors sent up here to run the border line got drunk and accidentally swerved miles north of their proper course. I'll bet you've heard that one before, eh? The fact of the matter is that they were all sober as judges. They were just plain wrong. What they were looking for was a stream draining into the lake from another system of lakes they supposed must lie somewhere to the northwest. They pushed up the backwater here by my house, thinking they'd found it. Of course, they hadn't. The 'inlet' of Angle Inlet is nothing but a blind bay that dead-ends a few miles back in the swamp. No matter. They slapped down the border here anyway. That was in 1872."

"When did you first come here?"

"In 1914, when I was two years old. I came up with my mother and father. My first memory is walking through the marsh grass right on the spot where we're sitting, and that grass was like a jungle, clear up to the old man's hind end. He burned it all off and built a cabin twenty feet wide by twenty-four feet long out of popple logs, and went to homesteading. After the Indians, my folks were the first people to homestead up here."

"What was this country like when you were a boy?"

"Sheer wilderness. A few Chippewa came here seasonally and that was all. We all lived off the land. Fished in the summer, hunted and trapped in the winter. Come spring, the fur packer showed up to buy our pelts. He was a great big giant of a French Canadian man, and there was a tale about him that he once carried six hundred pounds of fur on snowshoes all the way to Winnipeg. There was a real man for you. Nowadays, your antitrapping crybabies out in California have ruined the fur market so a fella can't make enough trapping to pay his taxes."

Joe Risser leaned forward and spat a stream of tobacco juice into his can: so much for the California crybabies. "Anyway, we lived a hardscrabble life. My mother grew a big garden. She kept a cow, and she and I picked berries. One summer she put up seven hundred quarts. We didn't have what you could call a town, of course. Our only neighbors back then were the Indians. They taught me how to fish. My first outside job was skinning bullheads for old Chief Black Hawk. He'd come in off the lake from netting fish and I'd skin 'em. I was twelve years old and that was my full-time job.

"Then a few sport fishermen commenced venturing up here summers. I started guiding at fifteen. From 1929 to 1950 I guided muskie fishermen summers and trapped winters. Every day I rowed a wooden boat fifteen miles along the shore just off the inlet here."

Joe glanced at his big knuckles, and for a moment I could see his strong, gnarled fingers wrapped around a heavy pair of oars. "Guiding those fellas was a big responsibility. Lake of the Woods is very shallow and windy, a bad combination that makes for big waves. What's more, the wind can spring up before you know it, and from any direction. This lake gets rough quicker than any other body of water I've ever seen. I had to keep an eye on the boat, the sports, the bottom features, and the weather. The minute it started to blow, I was off the water."

"What did you do after 1950?"

"Built log houses. I built every log house in the Angle, and they'll be here long after their present owners if I do say so myself. Then I got itchy feet and went up to Alaska and built log houses there. Drifted back here to the Angle. Commenced making wooden furniture, which is how I make my livelihood now. Care to walk out to the shop?"

The shop was cluttered with lumber, woodworking tools, machinery, and firewood for another potbelly stove. Here, in his ninth decade, Joe Risser still put in a full workweek turning out children's furniture, sturdy chairs and tables along the lines of the Adirondack furniture popular at summer camps before the turn of the century. "Kids love 'em," he said. "They're like my log houses. They last. In the late winter, before the sap rises, I go out to the woods and cut my timber, red oak and black ash. Then I saw up my boards and put them under cover to cure. These chairs and tables are all glued and doweled. Not one nail in them anywhere. Sometimes I'll build someone an ice-fishing shanty, and in the wintertime I read and visit friends. I feed the deer when the snow gets so deep they can't travel. Mainly I work. My work's my recreation, you could say."

On the way back to my car Joe showed me several piles of

lumber seasoning in an adjacent three-sided shed. "One year for red oak to cure, two for black ash," he explained.

"I notice you've got a lot of black ash. You must plan to stay in business for a while to come."

"I do," Joe said. "I don't make all that much money, but I like my work, and so far I've been able to keep out of the hole and stay on up here. That's good enough for me."

Notes from the Red River Valley

By the end of the second day after we left Pembina, we had not a mouthful to eat, and were beginning to be hungry. When we laid down in our camp at night, and put our ears close to the ground, we could hear the tramp of buffaloes, but when we sat up we could hear nothing, and on the following morning nothing could be seen of them, though we could command a very extensive view of the prairie. . . . We started early and rode some hours before we could begin to see them, and when we first discovered the margin of the herd, it must have been at least ten miles distant. It was like a black line drawn along the edge of the sky, or a low shore seen across a lake.

— John Tanner, *The Falcon*

NOON. The Manitoba-Minnesota boundary. Immediately after crossing back into Minnesota, I turn due west. From here the border is a straight shot along the storied forty-ninth parallel across the remote High Plains of North Dakota to Montana, over the Rockies, all the way to the Pacific.

1:00 P.M. And here's Badger, Minnesota, the self-proclaimed mallard duck capital of the world, where a wild mallard or two seems to float on every roadside pothole, the brilliant emerald heads of the drakes gleaming in the sunshine. Over fields flat as flat soar long-winged marsh hawks, patrolling the drainage ditches for frogs and mice. This ever-flattening country is hawk heaven. Hundreds of feet above me, red-tails catch the thermals, riding the currents away up there with scarcely a dip of their wings. Rough-legged hawks perch on fence posts near the sparkling blue potholes, looking for an unwary mallard to pick off. Diminutive sparrow hawks swoop off telephone wires to nab roadside grasshoppers. Before I know it, I've left the woods entirely and entered the oceanic vastness of the plains.

* * *

2:30 P.M. The highway is nearly empty this Sunday before Labor Day. All other travelers seem to have reached their holiday destinations, and the hawks and mallards and I have the fields and sky and potholes to ourselves. Here on the eastern fringe of the great rolling prairieland, the maples and birches and thick swampy underbrush give way to gray-green cottonwoods, wheat fields that stretch on for miles, and acre upon acre of sunflowers — all nodding my way in the perpetual light wind that sweeps over the Great Plains night and day, three hundred and sixty-five days a year. Then, just ahead, I spot a double line of cottonwoods marking the banks of the Red River at Pembina, North Dakota. Without quite realizing that it would happen so suddenly, I've reached the midpoint of my trip through America's North Country.

3:30 P.M. Pembina. At eighty-two, Hugh Chambers still remembers a few of the very last Red River buffalo hunters. A retired farmer, Hugh works as a volunteer at Pembina's historical museum, where he shows me an authentic two-wheeled Red River buffalo cart: "These rigs could carry five hundred pounds of buffalo hides and meat, and a small horse could easily pull one fifty miles in a single day. They were completely watertight. You could remove the wheels and paddle one across a river like a flat-bottomed boat. The wheels were made from a single piece of wood cut off the butt end of a cottonwood tree. There wasn't a scrap of metal in a Red River cart, not one scrap, and when several hundred of these contraptions heaved into sight out on the prairie, why, the screaming of the wooden wheels on the wooden axles could be heard ten miles away. The old drovers I knew said it sounded like ten thousand pigs being butchered. Drovers had to plug their ears with their own chewing tobacco, or they'd go deaf or mad or both, listening to it. In Pembina's heyday, a million or more buffalo hides a year came through here in Red River carts."

3:45 P.M. "I've never understood how they could kill so many buffaloes in so short a time," I say to Hugh Chambers.

"From everything I've read, buffaloes were skittish and hard to approach."

"Here's how," he says, handing me a hefty, long-barreled gun. "This is a fifty-caliber Sharps buffalo rifle. It was invented in the mid-eighteen-fifties and weighs ten pounds. That's about twice as much as a modern rifle, but it was considerably lighter than the old black-powder guns. It fired a bullet three and three-quarter inches long. The saying was that with a Sharps rifle you could shoot a buffalo as far away as you could see one. I don't know about that. I do know you could shoot accurately up to six hundred yards. That was far enough. As to why they killed so many, pure shortsighted greed is your answer. Sometimes all they used was the tongue, which was considered a great delicacy. Sometimes just the hide. That they'd tan and make leather from. But if your grandfather went courting your grandmother with a horse and sleigh, he very probably had a buffalo lap robe like this one to keep warm with. It's shaggy and it isn't terribly handsome. But it kept that old buffalo warm at fifty degrees below zero and it would keep your grandfather and grandmother warm, too."

"The Brat Is Back"

Out here, after you finish school, there are three things you can do for recreation: hunting, fishing, and racing. I've never cared much for hunting and fishing, so that left racing.

— Mike Tomlinson, defending track champion,
Pembina stock car races

On the outskirts of Pembina, the cottonwoods and poplars I'd seen growing in people's yards and clumped along the river thinned out to a few straggling loners. Again I felt the thrill of coming into new country — coming into plains still rutted in places with the deep tracks of the screeching Red River buffalo carts! Around six in the evening, a mile or so west of town, I heard a terrific whining from somewhere back off the highway, rising and falling and rising again. I'd been thinking about the unearthly shrieks of those buffalo carts, and for a fleeting moment I half-wondered if my daydreaming had me imagining that I was actually hearing the damn things. But no, there was that high-pitched whining again. It was coming from a small fairground, where fifty or so stock cars were gathered near a dirt racetrack in front of a wooden grandstand. Without further ado I turned off the main road and headed over to join the fun.

The drivers hailed from Grand Forks and Minot, North Dakota, from Kenora, in western Ontario, from Thief River Falls, Minnesota, and Winnipeg, seventy miles north on the Canadian Red River. Their cars were painted in splashy yellows and reds and sky blues and emblazoned with the names of their sponsors: Detour Bar, Iron Door Saloon, Pembina Farmers' Elevator. A glance down through the drivers' names on the program was a peek at the genealogy of the area's settlers: Norwegian farmers, French Canadian voyageurs, Scottish Hudson's Bay post factors, with surnames like Halvorson, Harstad, McBain, DeLisle, Nostdahl, Braudette, and Gendron. There was even a

Terrance Blacklance, whose local ancestry stretched back even further.

As I entered the infield, the night's hopefuls were crowded together testing their engines while the pit crews called out instructions. The drivers were dressed in vivid fireproof racing suits and shoes. Their crews and wives and sisters and girlfriends sported racing jackets plastered with decals and embroidered with the names of their cars and men.

I'd arrived during the preliminary "hot laps" or practice laps, when at more or less random intervals five or six cars roar out onto the track to pack down the loose dirt before the main events. To watch the hot laps from the nearest possible vantage point, young men and teenage boys were standing on ten-foot-high stacks of worn-out tractor tires along the home stretch. I climbed atop one of these tire cones beside three local guys and stood there in the fumes and dust while chunks of black mud thrown up by the cars rained down on us. Clods of rich North Dakota earth, the same earth that grew flax and sunflowers and wheat and corn as high as the eyes of the mastodons that once roamed these plains fell all around us, spattering our jeans and work boots and hands and faces. The flying mud was dark as night and as fertile as any soil in the world. It occurred to me, standing above the track as the drivers skidded nearly into our laps before gunning their multicolored junkers into the home stretch, that even on a holiday, the grinning young men sipping Budweiser out of cans beside me couldn't get away from the black earth that had given them and their ancestors a livelihood.

Suddenly a car swerved out of its lane and banged into another one. The colliding vehicles locked together, skidded sideways, separated and continued on their course. No apparent damage done, but "Hey!" a tall man nearby exclaimed angrily. "There's no call for that in a hot lap!"

As I jumped down off the tires, he turned to me and said, "That guy tried to rub Julie out before the race even begins."

Julie?

While the hot-lap cars rumbled off the track, I headed back

into the infield beside the indignant spectator. The car that had been bumped, Number 93, stuttered up beside us. Across the crumpled trunk the logo THE BRAT IS BACK was painted in large red letters. Out of the driver's window — stock-car doors are permanently bolted shut for safety, and there's no glass in the doors or windshield — climbed a slim person in a black racing suit and a bright red helmet. When the driver removed the helmet, a cascade of long blond hair fell over her shoulders.

It was Julie Osowski, the only woman contestant in tonight's races, who'd just turned eighteen and who, as of tonight, ranked third for the season in Pembina's Street Stox category.

"You okay?" the tall man said.

"I'm fine, Dad," Julie said. "No problem."

"There's no need for that sort of thing, either," her father said. "Here we go again. They just won't let up, will they?"

I introduced myself, and Julie smiled and said, "What Dad means is, as the only woman driver on the Red River circuit, I have to keep proving myself. Even in the practice laps. That guy who smacked into me is in my upcoming heat. He's trying to psych me out."

"It doesn't look to me as though he succeeded."

"No way! I've been racing for two years, and the intimidation days are over. That sort of stuff doesn't bother me at all anymore. It can't, if you want to win. Hey, my heat's coming up. I'll catch you afterward, okay?"

Julie crawled back through the window into Number 93, and her father, Dean Osowski, invited me to watch the race from the cab roof of the flatbed truck he uses to haul her car. But before the rumbling Street Stox moved out onto the track, Dean ran back up to Julie's car and leaned in through the window, presumably to give her a last-minute tip. When he returned I asked him if watching his daughter race made him nervous.

"No, it's actually very safe. We have far more farming and highway accidents around here than injuries on the track. What concerns me more is it looks like that car that tried to take Julie out in the hot lap is an outlaw Canadian."

"I thought Canadians could race here legally."

"They can. Half the cars in Julie's heat are being driven by Canadians. But I'm guessing that guy's car is five hundred pounds too light and just that much faster."

Seven cars, including 93, proceeded onto the track from the infield. They made a slow lap, lining up in prearranged positions. The green flag went up and down, the volume of the car engines increased exponentially, like the sound of a jet plane on its liftoff run, and within one lap Julie, who'd started fourth, was out in front of the pack. As she took the lead a roar went up from the hometown grandstand. Then she was slaloming around the curves in a flying deluge of black dirt and gunning "The Brat Is Back" down the straightaways just as hard as it could go, maintaining a three-car lead until the last lap, when the outlaw car from Manitoba started creeping up. "Don't give space, Julie, don't give him any space," her father exhorted her, just as if she could hear him; and although she didn't give an inch, and the Canadian couldn't take her lane away from her, he crossed the finish line first, by half a length.

A grim-faced Dean Osowski was waiting for the Manitoba driver with a judge when he rolled into the infield, and while they went off to weigh the suspect vehicle, I asked Julie what it took to be a successful stock-car racer.

"Concentration," she said. "You have to be able to focus completely on your driving for the duration of the race. It doesn't sound all that hard because you're only out there for a couple of minutes. But you'd be surprised how demanding it is to concentrate one hundred percent of your attention on the race for even that short time. A lot can happen in just a second or two, and when it does you can't forget two rules Dad taught me when I first started. Never give space and stay in your game plan."

"Were you scared when you first began racing?"

"Nervous, sure. Never really scared. I have pretty good reflexes and judgment, and best of all I've got a great teacher in Dad."

Just then the announcer reported over the PA system that the

Canadian had been disqualified for racing an underweight car, and the winner of the first heat was Julie Osowski.

"What now?" I asked her when the thunderous applause died down. "Turn pro?"

Julie laughed and shook her head, her hair glowing as golden as the surrounding wheat fields in the setting sun. "As of day after tomorrow, I'll be starting as a freshman at the University of North Dakota. Racing is just a hobby for me. Nothing more."

Julie and I shook hands, and I wished her good luck. On my way back to the car I met Dean, returning to his truck from the scales. "Good going!" I said.

"I knew the only way that guy could beat Julie was if he was underweight. He tried to hide it by sneaking up on her gradually. Forget it. Come back and see us again in a year or two. By then Julie'll be track champion."

Entering the Plains

Away to the Great Plains of America, to that immense Western short-grass prairie.

— Ian Frazier, opening sentence of *Great Plains*

Half an hour later, at sunset, I ask a tiny, harried-looking woman with a gaggle of young kids, coming out of a dilapidated store in the middle of nowhere: "Could you please tell me how far it is from here to Rolla?"

"Further than heck," she says cheerily. "Everyplace out here's a heck of a long ways from everyplace else."

West of Pembina the prairie is full of potholes. Potholes no larger than a mud puddle, potholes the size of stock ponds, potholes covering a few acres. Over all this water and grass, the Great Plains night comes with startling suddenness, the prairie seeming to exhale darkness the way the potholes exhale mist, though the sky itself stays light much longer.

Onward into the settling dusk, past bait shops on the porches of one-story houses, abandoned one-room wooden schoolhouses collapsing into dirt playgrounds overrun with wild sunflowers, boarded-up railroad depots, and little towns discernible five miles away in the twilight by their soaring white grain elevators. Past wooden water towers looming up on spraddled legs and towns buttoned down for the night at eight o'clock, on across the prairie, bisected somewhere nearby by an invisible border.

Another Side of the Kingdom

I killed a deer over there under that maple tree one night last year. With a machine gun.

— A student in my Northeast Kingdom adult basic literacy class

Meeting Julie Osowski at the Pembina stock-car races reminded me how much I'd enjoyed my early teaching years. They had provided me with the best imaginable introduction to Vermont's Northeast Kingdom. Through fishing and hunting with my students and getting to know their families, I accumulated a treasure of material in a very short while. I enjoyed teaching for its own sake as well. What a splendid forum it was for a storyteller — but in time I began to think that I liked teaching too much for my own good as an aspiring writer. I was spending every waking minute at work, leaving me little time or energy for writing. I realized that I had to get out of the classroom.

This was the late sixties, a time when many local young people were dropping out of the system and opting to live at one of the several communes that had recently sprung up in northern Vermont. For my part, I didn't want to become an expatriate from the community where we'd put down stakes, and neither did Phillis. Instead we stayed where we were, while I hired out on farms, clerked at stores, wrote feature stories for local newspapers. One of my best jobs during this time did involve teaching, but of a very different kind. I landed a job as instructor of a basic literacy class of twelve low-income adults, in the basement of a small church within sight of the Canadian border. At first, teaching reading and writing to grownups (some twice my age) was rough sledding. But when, partly in desperation, I hit on the idea of helping the class members write their autobiographies, the course took off. Most of the unemployed loggers, truck drivers, waitresses, mill workers, and farm hands in the class could read and write a little; but often I had to personally transcribe their drafts, and in a few cases I had them dictate their stories to me. What rich stories they were, and they seemed eager to share their lives with me and with each other.

Zeke, an out-of-work pulpwood cutter in his forties, was so accustomed to sitting on a log or stump out in the woods that he recounted his stories to me squatting on his heels in the corner of the room. Marie, a seamstress who had come over the border from Canada three years earlier knowing not ten words of English, wrote a compelling two-hundred-page memoir during the sixteen-week class, then went on to earn her high school equivalency diploma and enroll at the local community college. Elmore's father had sold him to a neighboring farmer at the age of ten, and he'd been kept virtually as a slave until he was sixteen. Percy, who preferred to be called Hound Dog, had been on the road since he was thirteen, slept in barns and town door ways, and prided himself on his status as an outcast. All these people had heart-wrenching stories to tell, stories that had never been told before and that I never would have heard in a high school classroom.

Something else was going on during the late 1960s and early 1970s and that, of course, was the growing opposition to the war in Vietnam. This opposition came rather late to the traditionally conservative North Country of Vermont. Both Phillis and I opposed the war vigorously and publicly from the start, so for several years I couldn't have returned to teaching in the local school system even if I'd wanted to. Phillis left public school teaching, too, to work at a rural sheltered workshop for mentally and emotionally handicapped kids and grownups. In the meantime I'd signed on with the Orleans County Council of Social Agencies, widely recognized as one of the most comprehensive and innovative antipoverty programs in the country. Under the brilliant and flamboyant direction of my close friend Tom Hahn, OCCSA sponsored a tremendous range of local programs in housing, transportation, education, job training, and health. My job was to do general social work with teenagers who had dropped out of school.

Many of the kids Phillis and I worked with had no place to live and eventually wound up staying with us in our ramshackle farmhouse. A girl named Bev showed up one January night during a blizzard, bleeding from a bad cut on her hand. "Me and my mother just had a knife fight," she proudly announced.

"Your mother attacked you with a knife?"

"Na. We had a knife *fight*. She cut me and I cut her. I guess I'll stay here with you guys for a few days, you don't mind." (She stayed for a year.)

Gail, who at sixteen drank a fifth of booze every day, appeared on our doorstep one summer morning with the intention of drying out. She did just that, stayed on two years, and graduated from high school. Alan, a speed freak, lived with us for a year. He wasn't as lucky as Gail. He killed himself on an overdose in Florida about the time she entered college.

Maggie had come to the Kingdom from California as the fifteen-year-old mail-order bride of a local ne'er-do-well three times her age. She would later become the model for Claire LaRiviere, the murdered French Canadian wayfarer in *A Stranger in the Kingdom*, though at the time I met Maggie I was still years away from writing that book.

Henry had gotten frostbite on his right foot at Earth People's Park, a border commune which for a time in the late sixties was a mecca for outlaws on the lam from all over the country. All Henry was able to tell me about his background was that he'd once "lived in a Vermont town with a river running through it." He lived with us for a time while teaching himself to read and write. We helped him locate his mother, who was as proud of his newly acquired literacy as any parent of a Harvard Ph.D. Then one night, just as Henry was about to leave for the Job Corps, he attended a party at a nearby lake and was never heard from or seen again.

Inspired by the work I had done and the people I'd met as a teacher and social worker, I laid aside my faltering on-the-road novel and began work in earnest on a series of Northeast Kingdom stories, which eventually found their way into my collection *Where the Rivers Flow North*. Yet in much the same way that I'd needed to come to terms with the Kingdom's darker side, I had to leave it for a time and then return to achieve the perspective to write about it clearly.

The Veterinarian and the Visionary

The history of the Métis community . . . is a tragic example of how the gestational process of a national consciousness can be aborted.

— Gilles Martel, *French America*

Bottineau's a great place to live, a great place to raise kids, but making a livelihood in and from the northern Great Plains imposes its own imperatives on us. When your wheat's ripe, you have to harvest it. If you're a hog trucker, you make your runs when the hogs are ready to go to market. If you're a vet, you get in the car and go whenever an animal's sick or down.

— Dr. Anne Scully, DMV, Bottineau, North Dakota

Dawn at the Turtle Mountain Cemetery. The neat, well-kept plots were still covered by frost, and a light glaze of frost lay on the glass display cases containing statues of Mary, bright plastic flowers, photographs of the deceased, and memorial plaques. Small American flags flew from many graves, with the dates of death coinciding with the dates of World Wars I and II, Korea, and Vietnam. Each plot was scented by wild sage, sharp and aromatic on the cold morning air, and the names on the stones were familiar to me. Trottier, LaVallie, LaFontaine, Grenier, DuBois — the same names I'd found in the cemeteries of the Madawaska Republic of Maine and the rough old overgrown boot hill on the outskirts of Seney, in Michigan's Upper Peninsula, and the lumber-town burial places of northern Minnesota. But the names on the stones of the Turtle Mountain Cemetery, near the North Dakota–Manitoba border, this frosty morning belonged to the *métis* descendants of French voyageurs who, like their Acadian countrymen back East, were for many years a dispossessed and displaced people.

After the explorations of the Red River area during the mid-eighteenth century, the North West Fur Company was founded in competition with the Hudson's Bay Company. From the start there were significant differences between these two great North American fur-trading houses. In contrast to the Hudson's Bay policy of establishing permanent trading posts where Native Americans came to sell furs, the North West Company traders aggressively ventured out into the bush to the Indians. And unlike Hudson's Bay, which hired Scottish or English factors to manage their posts, the North West Company interpreters, guides, and clerks were mostly French. Many married Native American women — thus the term *métis,* meaning half-breed — and established families in the Red River Valley. Inevitably, the trading territories of the two companies overlapped, and in the early nineteenth century a bloody all-out war broke out between them.

By 1820 the wealthier Hudson's Bay Company had emerged victorious, leaving about six thousand disenfranchised French-speaking *métis* scattered through the Red River Valley, on both sides of the border, to eke out a living from subsistence farming, supplemented by spring and fall buffalo hunts. Though they weren't rounded up and deported like the Acadians, the *métis* neither belonged to nor acknowledged any government other than their own. Their first communities, in fact, were nothing more than loosely organized hunting camps, which their spokesman, Louis Riel, described as follows in his memoirs:

> The *Métis* had almost no government. Nevertheless, when they went hunting bison, conflicts of interest naturally built up among them. As much to maintain order in the ranks as to guard against horse-thievery and enemy attacks by the Indians, they organized themselves and made up a camp. A leader was chosen, and twelve councillors elected . . . [who] established the regulations which were called the laws of the prairie.

The laws of the prairie worked remarkably well for the tiny *métis* buffalo-hunting camps and seasonal farming communities until 1869. In March of that year the Hudson's Bay Company transferred Rupert's Land, which consisted of the drainage basin of Hudson's Bay, including the Red River Valley, to the Confederation of Canada. The Canadian government immediately dispatched surveyors and soldiers to the new acquisition to establish sections of land to be granted to settlers from Ontario. The infuriated *métis*, who had been given no voice in the transfer of Rupert's Land, elected Louis Riel secretary of a national committee for independence; when Manitoba's official governor-elect arrived at the border north of Pembina to plant the Canadian flag on *métis* soil, Riel and a ragtag army of buffalo hunters, Red River cart drovers, scouts, fur traders, and prairie men unceremoniously turned him back.

Riel, who was educated as a priest, remains to this day one of the most interesting figures in the history of the North Country. For nearly a year he and his followers managed to maintain the Provisional Government of Rupert's Land — shades of the autonomous Indian Stream Republic, back in northern New Hampshire! — whose flag displayed a gold fleur-de-lis on a white background. When more than a thousand heavily armed British and Canadian troops showed up in Winnipeg to suppress the rebellion, Riel wisely chose to negotiate for the *métis* territory's entry into the Confederation. In 1885, however, he was hanged as a traitor for leading a similar rebellion farther west in Saskatchewan.

There was a fine exuberance this morning about the Bottineau border crossing north of the Turtle Mountain Cemetery, in the heart of the area where the Franco-American descendants of Father Louis Riel's *métis* insurrectionists today own and operate some of the most productive wheat farms in the world. Trucks headed into the States were lined up a dozen deep, with uniformed customs officers bustling from cab to cab to check hogs and beef cattle en route to Midwestern feedlots and slaughterhouses. The morning breeze was redolent with urine and manure, and through the slatted sides of the animal trailers came a medley of mooing, baying, bellowing, squeaking, and snorting. As I moseyed into the customs office, a slim, red-haired woman of about fifty passed me. She was carrying a bundle of documents, and she walked purposefully, with a little cowboy roll. She was dressed in jeans, a flannel shirt, and boots.

"Who's that, a special agent?" I asked a customs inspector.

"That," he said, "is the hardest-working person within a fifty-mile radius of Bottineau. She can outwork any man I've ever met, and here in this country where people still pride themselves on going full steam from before first light until long after dark, that's saying something. Her name is Dr. Anne Scully; she's our United States Department of Agriculture veterinarian."

Dr. Scully had deposited the documents on a desk and was headed outside again. Half-trotting to keep up, I asked if I could tag along after her. She kept right on walking fast toward the waiting livestock trucks. After maybe five seconds, she said, "Well, you seem to be."

Taking this as a yes, or at least not a no, I followed Anne Scully through a maze of eighteen-wheelers at a clip brisk enough to keep her heart-healthy for at least a century. As she checked the door seals, she told me she'd had a private veterinary practice in Bottineau for twenty-seven years, but she'd given it up three years ago to work full-time for the USDA as a means of "slowing down."

At the end of the line of hog trucks, a cattle carrier loaded with big slab-sided red and white Herefords was backed up to a pen. "These are Canadian steers going to the feedlots in St. Paul," Dr. Scully explained as we climbed over a wooden fence into a big pen. "I have to check them individually for tuberculosis and brucellosis inoculations. You may want to sit on the fence. It's going to get crowded down here in a minute."

As green to all this as a first-time rodeo spectator, I perched on the top rail of the fence, as instructed. "Okay, send 'em down," she told the trucker, and one right after another the colorful Herefords, raised in Canada for American supermarkets, came skidding down the urine-slick chute from the truck into the muddy pen while Dr. Scully checked their ears for inoculation tags.

On our way back to the customs building, it occurred to me

that this gracious, competent, hard-working woman could just as easily have been walking across campus to teach a seminar at, say, the Cornell School of Veterinary Medicine or on her way to a lucrative private vet practice in any one of ten thousand American suburbs. Why Bottineau, I asked her.

"Well, I was raised here and love the country. My family's been rooted in these parts a long time. I've always liked working with big animals, and after graduating from the University of North Dakota, I applied to vet school at Kansas. At the time, Kansas was one of the few schools of veterinary medicine in the country that accepted women. There were sixty-two men in my class and I was the only woman, but I got along fine after the guys saw that I could handle the big animals as well as they could."

Dr. Scully paused to watch the truck she'd just cleared rumble down toward the customs checkpoint. Then in the matter-of-fact manner I was becoming accustomed to along this stretch of the border, she said, "There's something else about small border-country communities like Bottineau that's very appealing. They really are still very good places to raise a family. I've got twin daughters, seniors in high school now, and this has been a wonderful place for them to grow up."

"What do your daughters want to do when they graduate?"

Anne Scully smiled. "One wants to go into premed. She's going to be a fine doctor, I'm sure of it. The other one loves animals. She'd be a great vet, but to tell you the truth, I'm afraid she's had too close a look at her mom's work schedule to want to be one. All the time my girls were growing up I was on call twenty-four hours a day, seven days a week. Like it or not, that's the reality of being a rural vet in Bottineau, North Dakota."

Del Haberman, a soft-spoken man in his middle fifties who manages the Bottineau Farmers' Co-op, had planned to take Labor Day off. Instead, he'd wound up coming to work early that morning, as usual, because with the run of good sunny drying days in a part of the North Country where weather exerts the same shaping force on the people as the ocean in Lubec and the big woods in

Madawaska, he knew that farmers from all over the region would be coming into town today with the first grain from their fields. When I arrived he was watching a big rust-red farm truck with Manitoba license plates unload. The *métis* driver, who looked and dressed like a cowboy but spoke with a French accent, pulled up onto a grate in the pavement, raised the bed of the truck, and released the tailgate. Tons of golden-brown barley cascaded down through the grate onto an underground conveyor that carried it up into a nearby elevator. When the truck was nearly empty, we went into an adjacent building, where a computerized scale was automatically recording the tonnage of the newly arrived barley in flashing red and green digits. Nearby the farmer's grown son, another French-speaking cowpoke, watched anxiously as a co-op worker tested the protein content of a handful of the grain in an X-ray machine the size and shape of a microwave oven.

Del Haberman watched too, with a concerned frown. "The farmers are all worried that the wet weather has lowered the protein count of their grain," he explained. "Thirteen percent protein, which is average, is worth three dollars a bushel."

Just then the figures 13.6 popped up on the gauge of the X-ray machine. A smile as wide as a Manitoba wheat field appeared on the young farmer's tanned face. "That's three-fifty a bushel for your barley, my friend," Del said. "Congratulations!"

It was too lovely a day to visit indoors so we went back out to watch another truck unload. I looked up at the soaring elevators. High above them, etched dark against the blue sky, a dozen or so very large birds were winging their way south in a straight line.

"Those are sandhill cranes," Del told me. "They migrate about this time of year. Aren't they beautiful to see?"

They were. And somehow the fact that they flew in a line rather than a V seemed in keeping with the straight lines on these northern plains: straight roads, straight railroads, a long straight border without a single jog in it all the way to the Continental Divide and beyond. The houses and barns and grain elevators were built with straight lines by straight-talking people who might at first appear to be two-dimensional themselves, in their straight, lean appearance and in their outlook as well, which seemed characterized by an absorption with work, family, their own immediate community, and an eternal preoccupation with the weather.

And then, right out of the blue, they surprise you.

I was more or less making conversation when I asked Del Haberman what he thought about the North American Free Trade Agreement, being implemented right in front of our eyes with all this Canadian grain pouring into American elevators. His answer, however, astonished me.

He began by echoing many another North Country resident's total dismissal of the U.S.-Canadian border as anything other than an imaginary line found only on maps and in musty old treaties. "When we first started taking Canadian grain, a few co-op members objected out of a sort of territorial instinct. Bottineau is so far away from any meaningful state or federal governmental entity, though, that for all practical purposes we might as well be in Manitoba. Whatever helps local farmers on both sides of the border is going to be good for the entire region in the long run. But the issue of who buys grain and where it goes runs much deeper than the relationship between Manitoba farmers and North Dakota farmers.

"This is how I see it," Del continued. "There's a worldwide need for wheat. In the undeveloped countries, especially, that need is desperate. But the capacity to pay for the grain simply doesn't exist in many of those same countries. In my view, the wealthier nations should subsidize a comprehensive global program to buy all the grain necessary to feed the poorer nations until they can get on their feet. There's absolutely no acceptable reason why anyone in today's world should ever be hungry, much less die of hunger. In time, underdeveloped countries could subsidize their own wheat imports to feed their people as a first priority. It all seems so simple, and really, it is. Think of the people we could keep from starving over the next months just with the half-million bushels of wheat in one of our elevators."

Del Haberman spoke so matter-of-factly that it took me a moment to realize that he was articulating a profound world vision. As he walked me to my car, he put it all in perspective in the same quiet manner. "You were asking about border communities," he said. "The point I'm making is that we're all part of a much larger community, a global community where people depend on each other to get by, just as we here in Bottineau do on a smaller scale. We all have to be part of that global community to make it work."

Notes from the North Dakota Prairie

I'm in quest of the land and what informs it.

— William Least Heat-Moon, *PrairyErth*

NOON. Like the deep North Woods of New England and the watery North Coast of the upper Midwest, the northern Great Plains along the forty-ninth parallel create their own distinct world. The radio station from Minot warns me to drive especially carefully today because combines and tractors and farm trucks are out on all the roads. SLOW FOR FARM EQUIPMENT signs appear every three or four miles, reminding me of the cautionary moose-in-the-road signs back in New England. Tiny sporting goods outlets tucked onto the front and side porches of houses proclaim AMMO and ALL BOW HUNTING EQUIPMENT HERE. Roadside eateries look more like classic western cafés, rambling one-story affairs with an appealing rough-and-ready aspect. Oil derricks hobble nervously in place in hay and wheat fields, sucking hydrocarbons up from beneath the rich black prairie earth to fuel the combines and tractors and the cars of middle-aged writers like me, now passing through such deliciously named North Dakota border towns as Antler, Loraine, and Bowbells.

1:00 P.M. The High Plains of North Dakota spread out and out, stretching clear to the horizon in every direction. The miles unroll. I sip a Coke and listen to country music out of Estevan, Saskatchewan. Pam Tillis and Tricia Yearwood and Garth Brooks are good company, singing genuine, down-to-earth songs, but since leaving home I've already heard the same songs half a dozen times apiece. Finally I snap off the radio and just drive.

3:00 P.M. In midafternoon I visit the Des Lacs National Wildlife Refuge, a haven for ducks and geese flying south from their Canadian nesting grounds to Texas and the Gulf of Mexico. The refuge, founded by Franklin Roosevelt in 1935, teems with migrating waterfowl in the spring and fall, but I'm a bit early to see them. Today the protected ponds and marshlands look as empty as the surrounding prairie.

3:30 P.M. Westward into coal country, where immense dredges scoop up tons of the stuff in one shovelful, lighting up at night like mobile skyscrapers inching their way over the prairie. All across the northern plains, these behemoths gouge into the land and strip away every last crumb of the rich black topsoil to get at the soft bituminous deposits lying just below, leaving a nightmarish desolation behind, worse even than the scalped clear-cuts of the forests of northern New England, because the topsoil, once gone or dozered off into spoil heaps, as slag piles are called, will never come back. This is, hands down, the bleakest stretch of North Country I've passed through. Once a tropical forest of palm trees and giant ferns, with dinosaurs that may, for all we know, have been as colorful as a High Plains sunset and as warm-blooded as us, it's all been reduced to a weed-strewn spoil heap for miles in every direction.

4:00 P.M. CANADIAN DOCTORS AND NURSES — NORTH DAKOTA NEEDS YOU, huge billboards on the edges of forlorn little coal towns implore. I'd seen similar pleas in northern Maine and Michigan's Upper Peninsula: PLEASE! OUR TOWN NEEDS A DOCTOR.

4:30 P.M. Several times this long afternoon I've slowed down to take a close look at armored trucks coming my way in what appears to be a widely spaced convoy. Where are the banks? And what are these dun-colored, steel-plated rigs doing up near North Dakota's remote border on Labor Day when the banks aren't open anyway?

Outside Lignite I turn onto an unmarked lane that I've been told about. It looks as though it might lead off to a coulee or a cattle watering trough or a hay or grain field or an abandoned Burlington Northern spur line or a disused lignite spoil heap. It doesn't, though. It ends near a fenced-in slab of concrete marked by a six-foot-tall silver spear designating the location of a subterranean chamber housing a nuclear weapon capable of annihilating every last soul in a Siberian city ten thousand miles away.

A Close Brush

This, finally, is the punch line of our two hundred years on the Great Plains: we trap out the beaver, subtract the Mandan, infect the Blackfeet and the Hidatsa and the Assiniboin . . . ship out the wheat, ship out the cattle; dig up the earth itself and burn it in power plants and send the power down the line; dismiss the small farmers, empty the little towns; drill the oil and natural gas and pipe it away; dry up the rivers and springs, deep-drill for irrigation water as the aquifer retreats. And in return we condense unimaginable amounts of treasure into weapons buried beneath the land which so much treasure came from — weapons for which our best hope might be that we will someday take them apart and throw them away, and for which our next-best hope certainly is that they remain humming away under the prairie, absorbing fear and maintenance, unused, forever.

— Ian Frazier, *Great Plains*

The fenced-in site at the end of the lane was no more than a couple of hundred yards off the highway, but there was a terrific feeling of remoteness here. As I got out of the car for a closer look, the utter emptiness of the Great Plains along the forty-ninth parallel hit me like a natural force, like snow or wind or cold or darkness. Except for the blowing grass, I didn't see a single living thing. Not a lark or hawk or prairie dog or mouse, though as I approached the fence, stopping fifteen or so feet away, I watched where I put my feet for fear of stepping on a rattlesnake. But there weren't any snakes here either. The only sound was a truck on the highway. A native of mountains and abrupt wooded hills and quick rivers, I was nonetheless deeply attracted to this austere place, whose apparent barrenness of life I knew to be an illusion. Yet I simply couldn't imagine living, much less farming or ranching, near a missile site whose exact geographical coordinates right down to the precise seconds of latitude and longitude were beyond all doubt programmed into an identical missile in Russia, all ready to let fly at a moment's notice. A sign on the fence proclaimed that it was unlawful to enter the enclosed area and that the "use of deadly force was authorized" on anyone who did. The irony made me laugh aloud — an out-of-place sound that enhanced the eerie emptiness of the spot.

As I started back to my car, still half-watching for snakes, an armored vehicle like the ones I'd been seeing out on the highway pulled in and headed slowly up the lane in my direction. I waved, and it slowed to a crawl, its tires crunching on the loose gravel as it crept closer. The driver, invisible to me behind a narrow tinted windshield, blinked his headlights on and off several times. I waved again and got back in my car.

The vehicle stopped altogether a hundred yards away. In my rear-view mirror I could see its lights continuing to flash. Unhurriedly, I backed my car into the turnaround space at the end of the lane and started out toward the armored truck, now blocking my path.

I was getting annoyed. This road wasn't posted, and I wasn't trying to make a run for it. What the hell was going on?

No matter, I had to stop or detour into the prairie around the truck. I thought about getting out of the car, but instead I rolled down my window, and after about ten seconds a heavyset guy in his midtwenties, wearing a camo uniform with sergeant's stripes on his sleeve, walked up to my car.

As he approached I said, "My name's Howard Mosher and I'm writing a book on the country along the border. I stopped here to take a few notes."

"Sir, this is a restricted air force area, sir," he said very loudly, considering he was standing less than a foot away from me.

"I read the sign on the fence and didn't go any closer," I said. "This road isn't posted. The restricted area is back there inside the fence."

What I said was indisputable. There were no posted signs on the gravel road or the adjacent prairie, and the young sergeant looked momentarily puzzled. Then he said, "Sir, I'm going to have to ask you to shut off your engine and stand by, sir." This time when he spoke, double-sirring me again, I could see that he was nervous, which made me nervous, too. More than that, I was angry about being detained by the U.S. Air Force on unposted property.

"Where are you going?" I said sharply as the sergeant started for the armored vehicle.

"I have to call command headquarters about this situation, sir."

"There isn't any situation. I'm a writer, with a ton of identification. Go ahead and call if you have to, but tell your headquarters people I want to talk to them."

A few minutes later the sergeant got back out of his vehicle. "Sir, you're lucky this time. You can proceed on about your business. Command headquarters said if you have any questions about the base, you can call the Minot information officer."

"Good. I have a question about a civilian being detained by the air force on unposted property."

"Well, you see, sir, if you touch that fence an alarm goes off. Then we could call the local sheriff out to escort you off the property."

"Yes, but I didn't touch the fence and you detained me anyway. I notice that you're authorized to use deadly force if someone gets inside the fence."

"Yes, sir, we are."

"Are you authorized to use deadly force up here on this unmarked lane?"

"You'd have to ask our information officer about that, sir. I don't really know for sure."

"Well, I'd hope to heaven you'd know before you threw down and shot somebody for being on unposted property."

I wasn't really angry at the young sergeant anymore, just at the air force (and the other branches of the service for good measure) for throwing their weight around and not posting their facilities if they didn't want people snooping.

"It's a hell of a responsibility, isn't it, sergeant?" I said. "Guarding these weapons that could blow cities like Minot off the face of the map in a heartbeat."

"Yes, sir. It surely is."

"Well, you're doing a good job."

"I thank you, sir."

"Thank you, sergeant. Thank you for not using deadly force on me. Or, so far, on anybody else. Including the Russians."

Now my friend the sergeant grinned, despite himself, and we parted on good terms. In the end, he really was just doing his job, and he was right not to laugh about it. Nothing at all about this place was a laughing matter.

Notes from the Medicine Line

Surveyors . . . need credit and remembrance for a job finished swiftly and efficiently — a job of immense importance. And though they have never struck anyone as glamorous enough to be written up in a western story, a young man in search of excitement in 1872 could have done worse than enlist with them.

— Wallace Stegner, *Wolf Willow*

5:00 P.M. On I press, shut of the United States Air Force, through towns perched on the High Plains like islands surrounded by open-pit coal mines, towns flanked by invisible underground missile installations, and a town named Fortuna, where a sun shower smelling fresh as a prairie spring morning has just blown over, bringing with it a triple rainbow that takes up a full third of the sky, its immense pink, lime, and lavender bands ending not in a pot of gold but, better yet, in a harvested wheat field on the edge of town. Three combines sit by the road, glistening green and hopeful beneath the fading rainbow as the late-afternoon sun emerges again to dry out the standing grain in an adjacent field for the next day's harvest.

5:30 P.M. At the Montana state line, the pavement takes on a bluish sheen from some new pigment in the crushed stones. The terrain is hillier. RANGE CATTLE, an official highway sign ahead of me warns, and suddenly I have to brake hard for a herd of Herefords dashing out of an arroyo. A few minutes later a dozen pinto ponies swirl along a steep hillside beside me for a few hundred feet, then disappear in one of the countless draws leading back into the hills. Montana! I'm as excited as I've been in a long time.

6:00 P.M. I swing off onto a gravel road leading up through northeastern Montana's incredibly jumbled network of cliffs and ravines known as the breaks, to the forty-ninth parallel and the

Canadian border. Here Marvin Crabtree, the western survey chief of the U.S. Division of the International Boundary Commission, is just loading a four-wheel all-terrain vehicle into his truck after finishing his day's work on a stretch of the border that hasn't, he tells me, been surveyed for eighty years. Back in 1981 Marvin and his crew reset the monument I'd stopped to look at on the Manitoba-Minnesota boundary of the Northwest Angle, after it had been shot to smithereens by hunters, and they've been leapfrogging their way west ever since, following in the footsteps of the Canadian and American surveyors who, from 1872 to 1874, pushed the international boundary (known to local Indians as the Medicine Line) westward through seventy-below-zero temperatures, clouds of mosquitoes, hailstorms with stones the size of hen's eggs, disease-ridden swamps, and summer heat so intense that for days on end it created mirages that brought all surveying to a halt except for a few minutes at dawn and sunset. Not to mention plagues of locusts, prairie fires, summer snowstorms, botflies, and hostile natives who'd as soon pick off a surveyor as one of the few buffaloes his white predecessors hadn't already slaughtered for hides or tongues or the sheer bloodthirsty fun of it.

"It doesn't look it, but we're about three thousand feet up," Marvin tells me. "We've got the original elevation figures from the 1874 survey to go by. Some of them are a bit off, but it's amazing how accurate those guys were. The instruments have improved, of course. But up here in the breaks, the country's essentially no different than it was more than a century ago when that first survey party went through and left stone cairns to mark the border."

7:30 P.M. Plentywood, Montana. This is a picture-perfect northern Montana town: a mile-long main street lined with one- and two-story buildings, rectilinear side streets lined with low cottonwoods, a short strip of motels, gas stations, and cafés on the west edge of town, and that's all she wrote, folks; that's Plentywood.

* * *

8:00 P.M. The Sherwood Inn in Plentywood. "What's a typical northern Montana meal consist of?" I ask the chef.

"Beef steak, beef steak, and beef steak," he says. "All local and all delicious. I'll cook you up the best charbroiled sirloin steak you ever laid a lip over. If it isn't, you don't have to pay a penny for it."

Meanwhile a hard-looking young woman in brief red shorts and a halter top, sitting at a nearby table with two guys who look more like carnival roustabouts than cowboys, starts shouting. "Look, I told you worthless sons of bitches I'd buy you a meal if you gave me a ride. So here's your goddamn meal. Anything on the menu. Order it and eat it and then get the hell out of here. A meal's all you get."

8:30 P.M. I tell the chef, Paul Overgaard, he was right about the steak, and he sits down at my table to visit for a few minutes. He seems to enjoy watching me eat the best beef steak I've ever laid a lip over. He's interested in my trip, impressed that I've traveled all the way out to Plentywood from the Green Mountains of Vermont. "My great-grandfather pioneered this area in the late nineteenth century," he says. "The first few years he wintered over in a cave out in the breaks. When I was a kid he took me there. It was just a hollow in the rocks, actually. I couldn't imagine anybody living in it, though at one time or another a lot of famous outlaws hid out up there. Butch Cassidy and the Sundance Kid. Sitting Bull. The Wild Bunch. We even had our own local bad men, like Outlaw Jack Van of Outlaw Coulee."

Outlaw Jack Van? My ears perk up.

"That's right," Paul Overgaard says. "You want to hear about Outlaw Jack, you slide over to John Olson's place tomorrow morning. John's eighty-nine, and he'll tell you all about Jack Van and Outlaw Coulee. John ranched most of his life up there."

The Ballad of John Olson

All northeastern Montana is hard country — the land is dry and sparse and the wind never stops blowing. The heat and thunderstorms in summer can be brutal, and the winters are legendary for the fierceness of their blizzards and the depths to which temperatures drop.

— Larry Watson, *Montana 1948*

John Olson lived in a bungalow on a side street near the high school, where I'd gone for a run the evening before, until the mosquitoes had driven me off; but with a rime of new frost on the sparse brown lawns and the sun just coming up over North Dakota to the east, there were no mosquitoes abroad in Plentywood this morning. I ducked into a rough-hewn vestibule full of tools, pails, lumber, and obscure automobile and truck parts, and knocked on the door, and after a minute a man who looked to be in his midsixties appeared in the doorway. At eighty-nine, John Olson had iron-gray hair, didn't carry an ounce of fat, and, as he told me a little later, still had every one of his own teeth, whose longevity he attributed to the preservative properties of the Copenhagen chewing tobacco he'd used regularly since he was sixteen. John invited me into his combination sitting room–bedroom and cleared a space for me on a straight-backed chair. He sat down on a rumpled daybed, positioning himself so that the rising sun shone in on his back and shoulders.

John's bungalow consisted of four rooms. From where I sat, I could see into each of them. There wasn't a speck of dirt or dust anyplace. The dishes and silverware were washed and stacked neatly on the kitchen counter. There was no food sitting around. Yet these rooms were as cluttered as any I'd ever seen, with the excess clutter from one room drifting over into the next. There were harnesses and horse collars, coffee cans full of nails and screws, bottles of glue, enough tools to stock a small ranch, summer clothes and winter clothes piled on chairs and old couches, photographs of horses and cattle and cowboys, scores of mahogany-colored canes and walking sticks with light swirls and checks and diamond patterns, fishing and hunting gear, automotive parts ranging in size from a thin red speedometer needle to the axle of a pickup truck, and, on shelves all the way around John's sitting room, brightly painted miniature stagecoaches, chuck wagons, buckboards, Conestogas, circus carts, and railway cars, many bedecked with blue ribbons.

When I told John that I was traveling through the American North looking for last frontiers, he gave me an appraising look and said, "My dad homesteaded just north of here, on the border, in 1901. It was still a *real* frontier up there then. You were apt to encounter about anything. Five-day blizzards, drought, half-starving Indians wandering around with no place to hang their hat, bad men on the lam, anything. This was wide-open country."

"Do you remember Jack Van of Outlaw Coulee?"

John grinned. "You bet I do. We lived in Outlaw Coulee. Jack Van lived the next place west from us. He was quite an hombre. The genuine article, you might call him."

"Was he really an outlaw?"

John chuckled. "You decide. Jack slid up to the breaks from Miles City, but some folks said he hailed from the Oklahoma Panhandle before that. Nobody knew what he was trying to get away from, if anything. He ran some slicks — unbranded horses — back and forth across the border. We never got involved with any of that, never asked any questions. The less you knew about Jack's business the better. Anyway, one day Jack went out with a hammer and a bucket of nails to fix his corral fence. He was working along, minding his own business, when up gallops this homesteader from over the line in Canada. Homesteader had a rifle, and he was raving mad. So Jack said afterwards, anyway. 'Tweren't any witnesses to back him up or contradict him either. According to Jack, the homesteader accused him of playing around with his wife and threw down on him with the rifle, and Jack flung the hammer at him and knocked the sodbuster bang dead off of his horse. There was an inquest down here in Plentywood, and Jack was cleared of any wrongdoing. The jury ruled self-defense after six minutes of deliberation.

"Now, mister," John said, warming to his story as the early sunshine warmed his back, "I went to the first school in Outlaw Coulee. Little one-room building where I learned to read and write. Kids from Montana and Canada both went to school there because back then you could live in Canada and never know it. We went to school three months in the spring and another month or two in the fall after the harvest. That was it for schooling. Summers and harvest time we were needed at home, and winters were too tough to keep school at all. When I turned twelve I left school to drive a grain wagon. After the harvest was in I got a job driving a coal wagon. That winter I drove a four-hitch coal sleigh. It carried three tons of coal, and I drove it fifteen miles a day from a little mine up by the border near our place down to the railhead. We had winter in those days. The snows covered the houses and sheds and shacks. But I had to drive that coal sleigh every day, blizzards or no. I'd like to live long enough to see one more real old-fashioned winter, just to see what people would do today to survive it."

"Do you think people up here were tougher back then?"

John thought about this. "Well, we were tough as rawhide, kids and all. We had to be. But I don't know that we were any tougher than folks today. The difference is, back then if you made up your mind to ranch, you could do it. It wasn't ever easy. But it was possible. Ranching was all I ever wanted to do, and when my dad lost our place to the bank, I was floored. We all were. They foreclosed on the tenth of June, 1922, the year I turned eighteen. So I said to the bank president, 'I'm going out to earn the money to buy back our place. You'll hear from me.' Then I left home and went to building grain elevators down along the Missouri River. On the fifteenth of August of that same year a fella who'd been hunting me for two weeks finally overtook me in Billings. He said the bank would sell the home place back to me on contract. They'd take half my crop each harvest until I'd paid off the price.

They couldn't find another buyer, you see; that's what accounted for their change of heart. Well, I had an old Ford Model A. I traded it for some heifers and some winter wheat seed, and I went home and went to work. Two years later I paid off the bank and took ownership of the place free and clear."

John nodded over that long-ago satisfaction, then said, "Now here's my point. Today a young man couldn't do that. A young fella couldn't make it starting out as a rancher with nothing but an old car to trade. I don't care how tough and determined he was. That's why so much of this territory out here is empty. Empty and emptying out. Look here. See that poem over there on the wall?"

John pointed at a framed poem next to some pen-and-ink drawings of cowboys with cattle and horses.

"Years ago I took in a kid and raised him from the time he was fourteen," John said. "When he got growed up he wrote that poem and sent it to me. He calls it a ballad. Go ahead and read it. It tells some of my story."

I got up and read the poem, which was titled, "Ballad to John Olson, the Man Who Raised Me." Part of it went:

> John had worked out some and saved and sharecropped.
> It was an uphill race.
> But he finally gathered up enough to buy back the old
> home place.
> He did a little trading and gathered up some cattle . . .
> He didn't owe the Federal Land Bank
> So when he got his cattle check,
> It was his to stay.

"Those drawings are the kid's, too," John said, and his eyes got a little watery. "He turned out good. Got a family of his own now and lives a couple hours from here. He still comes to visit whenever he can. He just needed a chance, somebody to give him a hand. He was a good kid. Now he's a good man."

For a minute neither of us said anything.

"That sun feels good," John said. "After eighty winters up on that ranch, the sun always feels good to me."

"Do you miss the ranch?"

"Yes, but I don't miss the isolation. Here I can see my town friends half a dozen times a day. That's good after being up in the Big Lonesome for eighty years. Ranch life is good, but the isolation part, that's bad."

"How about Jack Van? Did the isolation get to him in the end?"

"Hell, no. Jack was never alone that long. After that set-to with the Canadian homesteader and the hammer, he up and married the dead fella's widow, and so far as anybody ever knew they were very happy together."

I had to laugh and so did John. Then I told him I'd like to drive up to Outlaw Coulee and look around for Jack's place.

"Look away," John said cheerfully. "But you won't find much. All those old places are gone now, just open range and the breaks. And stories. Lots of stories. See Tim Clawson up there. He's got some good ones."

Just before I left, John Olson said to me, "Montana's been a great place to live. Still is, for that matter."

"What's the best thing about it?"

John thought for a minute. Then he looked right at me and said, "We still watch out for each other. Nobody steps on anybody else's toes. But we watch out for each other."

Jake Blodgett, Logger

Writing fiction is a risky business.

— Graduate student in the MFA creative writing program, University of California at Irvine, September 1969

There's nothing safe about working in the woods, schoolteacher. Watch your ass out there.

— Jake Blodgett, Northeast Kingdom horse logger and ex-whiskey runner, October 1969

After leaving John Olson's, I stopped a few miles north of Plentywood and walked out into the breaks to stretch my legs. Near the head of an old dry streambed I sat down on a cottonwood log to eat a snack of crackers and cheese. Somehow, eating outside in the Montana breaks, far from any town or city, and thinking about John Olson's friendship with the kid he'd raised, brought back to me in a flurry of images the fall of 1969, when I went to work in the woods for my Northeast Kingdom mentor, Jake Blodgett, the man my son is named for. The experience turned out to be the single most important turning point of my career.

To put that fall in context, I need to back up a few months to the previous spring, when I had what seemed like a good idea. With a few years of teaching and social work under my belt, desperate to find a publisher for my fiction — and, I now think, to find a shortcut to learning how to write a good novel — I enrolled in the MFA creative writing program for the next fall at the University of California at Irvine. Phillis and I left Vermont in late July, following much the same route I was now traveling, camping our way west, and arriving at Irvine just before the start of the fall semester.

On our way into town we encountered a man in a gorilla suit vigorously directing traffic at a busy intersection, which struck us both as funny, and, somehow, promising. We rented a three-room apartment in Santa Ana, and the next morning I ventured over to the university and drank coffee on an outdoor terrace with half a dozen other creative-writing graduate students. They were pleasant, serious people, who spent the better part of the morning speaking in grave tones about how hard it was to write.

The next morning I returned for more coffee with my colleagues. The conversation resumed where it had left off the day before. Again everyone agreed that writing was a difficult undertaking. It was hard to find the right subject matter, hard to find one's voice, hard to find a publisher. I liked the other students. I liked sitting with them in the California sunshine, sipping coffee and commiserating about how hard we all had it. But I had known long before coming to Irvine that writing fiction was heavy sledding. How many times did I need to be reminded?

On my way back to the apartment, at the gorilla intersection, a telephone company van pulled up beside my car. The driver stuck his head out of the window. "I saw your green license plate," he called to me. "I'm from Vermont, too. Go back home where you belong while you still can."

"Why?" I called back to him.

"Do it!" he yelled as he drove off, and somehow I knew instantly that he was right.

By midnight Phillis and I were in Las Vegas, posting east. Two and a half days later we crossed the Vermont state line. We had slightly over three hundred dollars to our name, no jobs, and no prospects of jobs. Still, we were home, where we belonged, and the following afternoon I found myself standing in the overgrown dooryard of Jake Blodgett's shack, a few miles south of the Canadian border, inquiring about a job as an apprentice horse logger.

I'd never met Jake before, but I'd heard plenty about him. One of the Kingdom's last frontiersmen, he'd reportedly made his living during Prohibition by smuggling whiskey over the border from Quebec; even now, in his late sixties, he was renowned throughout the Kingdom for an independence of mind bordering on misanthropy.

Jake met me on his listing doorstoop, a tall, well-set-up man with short white hair and sharp, pale blue eyes. He wore a clean, faded red flannel shirt, wool pants, and steel-toed logger's boots. About him, then and always, hung the pungent redolence of the fir and spruce trees he'd spent a lifetime cutting — a wonderful Christmasy fragrance, which, Phillis and I would soon discover, no number of washings can ever entirely remove from a woodsman's clothes.

I introduced myself and said that I'd heard he was looking for a man to help him in the woods. (I did not add that this was the *only* job opening I'd been able to scout up in the area.)

"May be," he said, making the single word into two to emphasize his skepticism. "What have you done before for work?"

I admitted — and it had the full flavor of an admission — that I'd been a teacher.

From inside the house came a hooting laugh. A bulky, white-haired woman about Jake's age appeared in the doorway: his long-time housekeeper-companion, Hazel. "A schoolteacher, Jake!" she said with delight. "You got a Christly schoolteacher wants to come to work in the woods for you."

"Yes, sir," Jake said grimly. "How much pay would you want?"

I shrugged. "I've never worked in the woods before. Try me out for a week and pay me whatever you think I'm worth."

"That wouldn't be much," Jake said.

Hazel whooped like a loon and bobbed her head with glee. "A Christly schoolteacher!" she said again. "Wants to work for Old Jake!"

Jake stared at me for a moment with those assessing blue eyes. Then he gave a short, bleak nod. "All right, Mr. Teacher. Be here tomorrow morning at seven."

"Come on in, mister," Hazel called to me when I presented myself at Jake's door the next morning. "Jake! New fella's here. Christly teacher's here. Showed up after all, he did."

Hazel thrust a mug of tea into my hands and sent me around to the rear of the house, where Jake was harnessing the woods horse inside its three-sided hovel. The horse, which was gray around the muzzle, refused at first to open its mouth for the bit. Jake grumbled and cursed bitterly, though neither then nor at any other time did I hear him raise his voice to the animal. I watched attentively, thinking that in a day or two I could harness the horse myself. In fact, it would take me (an ex-schoolteacher!) a full week to master the job.

The woods where we would be working began at the far end of a long stony pasture behind Jake's barn. Later he would claim that you could cross the pasture by stepping from rock to rock without setting foot on grass the entire way. As we drew nearer to the trees, the narrow logging road opened up ahead of us, exuding an aroma of evergreens and ferns and last year's decaying leaves. Already a few soft maples along the edge of the pasture were turning scarlet. It was beautiful.

My job, as Jake put it, was to run the horse, which skidded the logs he cut out to a clearing near the edge of the woods. Running the horse did not turn out to be an especially difficult job, but it required constant alertness. Around midmorning I took a break to get a drink from a nearby brook. As I was returning, Jake shouted at me over his idling chainsaw, "Where's your horse, schoolteacher?"

To my alarm, the animal was nowhere in sight. I took off down the skid path at a dead run, coming in sight of the absconding rascal about halfway to the clearing. Not wanting it to panic and break into a gallop, I slowed to a trot. The horse glanced back over its shoulder and mended its pace accordingly. When I speeded up, it did, too. Fortunately, the dragging whiffletree caught on a low stump at the edge of the pasture, and I was able to overtake the runaway.

This little charade, which tickled Jake no end, typified the rest of my first day in the woods. It was far from smooth, certainly, but somewhat better than I'd anticipated. At any rate, I got through it, and Jake didn't tell me not to come back the next morning.

"You smell good, like the woods," Phillis told me when I got home that evening.

Surprisingly, I felt good, too. True, every joint ached and I was stone tired. But for the first time since we'd left California, I had no misgivings about my decision to return to Vermont.

For all his alleged contempt for schoolteachers, Jake Blodgett turned out to be a born teacher himself. He never told me anything he could show me, but he let me make my own mistakes, without looking over my shoulder to see if I was doing things correctly. If the massive chain snubbed around the logs I was skidding pulled loose halfway to the landing yard, it was up to me to extricate myself from the pickle. At the same time, Jake clearly enjoyed passing on skills and lore to an apprentice; he took considerable pride in pointing out how little damage horse logging caused to his beloved woods, compared to the havoc being wreaked by the ten-ton gasoline-powered skidders that were turning whole mountainsides into wastelands elsewhere in northern New England.

My best times in the woods with Jake were our lunch hours. We had an unusually long run of warm, clear days that fall, and from twelve to one we'd sit in the sun surrounded by evergreen slash and sawdust from felled trees, eating our sandwiches while Jake yarned on and on about his life, telling stories that (though I didn't know this at the time) I would later use in my novels — stories of logging and hill farming and whiskey running in the Kingdom.

Two full months passed in this way, quickly and happily, and then, just as I was becoming proficient enough in my job as journeyman horse logger to earn some part of the eight dollars a day that Jake faithfully paid me, he seemed to decide that I had other lives to lead and should get on with them. One noon I had casually mentioned to him that I was interested in writing. Since he said nothing in response, I assumed that he hadn't given the matter a passing thought. But a couple of weeks later, on our way out of the woods together during the first snowstorm of the year, he glanced over at me through the thickly falling flakes and asked out of the blue what sort of writing I did.

I told him that for the past several years I'd been writing stories about the Kingdom, and though I hadn't had any luck publishing them yet, I intended to write more.

His pale blue eyes narrowed. "Would you ever write a story about me?"

"Yes. I'd write about your whiskey-running days. And your horse logging."

Jake gave a short nod and continued on through the snow. Then, just before we reached his place, he turned to me and used my first name for the first time. "Well, Howard," he said, "you'd better get on with it then."

Part Four

THE OUTLAW TRAIL

Outlaw Coulee

When cliffs, sheer drops under impossible overhangs, ended the road, the workers filled the ravines or built bridges over them. They climbed above the site for tunnel or bridge and lowered one another down in wicker baskets made stronger by the lucky words they had painted on four sides . . . Ah Goong had been lowered to the bottom of a ravine, which had to be cleared for the base of a trestle, when a man fell and he saw his face. He had not died of shock before hitting bottom. His hands were grabbing at air. His stomach and groin must have felt the fall all the way down.

— Maxine Hong Kingston, *China Men*

OUTLAW COULEE the sign said, with an arrow that pointed down a gravel road just south of the Saskatchewan border and parallel to it. Tim Clawson's ranch was a mile to the east, on a rise overlooking the coulee, a dry streambed at the bottom of a steep-walled cut. Beyond the unpainted ranch house was a corral with a few horses, and then, clumping into the hillside, several pole sheds containing elderly-looking farm machinery. SALESMEN WILL BE PROSECUTED, TRESPASSERS SHOT a notice on the door announced, but I couldn't imagine a salesman stopping at the Clawson ranch in the first place, and with John Olson's blessing to call on Tim, I didn't feel like a trespasser. A racket broke out inside when I knocked. Dogs barked, somebody shouted at them to be quiet, and after a minute a man who looked to be John's contemporary opened the door.

"I'm not selling anything," I said. "Are you Mr. Clawson?"

"Yes, sir."

I told Tim Clawson who I was and mentioned John Olson's name, and Tim said not to pay any attention to the sign on the door, he hadn't shot anybody lately. He seemed glad to have a visitor. "Come in, come in," he said. "We'll have some cowboy coffee."

Over black cowboy coffee "strong enough to float a horseshoe,"

Tim told me that he'd grown up right here on this ranch in Outlaw Coulee and was still hanging on, though just barely. "The fact is I'm down to operating the stock inspection pens at the border entry. I still get over there once a day, though I've got a man to tend to most of the upkeep work for me. How is old John, anyway? I haven't been down to see him or he up here in a couple of weeks."

"John's good. I saw him just this morning. He called this country up here the Big Lonesome."

"Well, that's exactly right. And John always did like company. Anybody does except maybe a hermit. But it's the winters I mind these days. They're just too long, you don't get out and see anybody for days on end. I can't have an interesting conversation with my dogs and cats, you know."

"Who's the most interesting person you ever knew up here in Outlaw Coulee? Jack Van?"

Tim Clawson thought about this while he poured us more cowboy coffee. "I don't know that I ever thought of Jack Van as particularly interesting," he said. "Jack was a hard man and probably on the wrong side of the law off and on. My family never neighbored with him the way we did with John's people. Oh, we'd help him out if he got sick or hurt, same as he would us, but otherwise we kept our distance. No, I'd say that the most interesting fella I ever knew in these parts was an old Chinese man named Poi Wing. Poi Wing had a general store up over the border in Big Beaver, and when I was a young shaver I hired out to work for him. Poi could speak English fine, same as you and me. But for some reason he liked to talk Chinese to me. Why, I don't know. Just to parlay in his native tongue, I reckon. It got so I could understand him some, enough to get along, and after a while that's all we spoke. We were friends, too. Not just a boss and a young kid working for him. Friends. Now that's an unlikely friendship, an old Chinese man and a young kid."

Tim grinned at this memory, and his eyes were a bit watery, too, like John's when he told me about the kid he'd brought up. "Old Poi raised some big white turkeys out back of his store. Come Thanksgiving, he'd slaughter 'em and wrap 'em in newspapers and pack up several dozen in the bottom of grain barrels, and then he and I would smuggle those turkeys over the border to Plentywood in a mule wagon. We'd laugh the whole way down and back. Thought we were getting away with something. Poi was very highly thought of up around Big Beaver as a shrewd, honest businessman, and I learned a lot from him. He'd had a hard life, left his family behind in China and came over here to lay track for the Great Northern for slave wages, saving every penny so he could buy that store. Nothing easy about it. But he never lost his sense of fun. Life was hard but it was fun. That's one thing he taught me. Those turkey-smuggling trips were some of the best times I ever had. You want me to drive you down to the coulee now, show you where Jack Van lived?"

Tim didn't bother to lock his door when we left. He said we wouldn't be gone that long, and the day he had to lock his door would be a sorry day in Outlaw Coulee. We rode in his farm truck, which he drove slowly down to the dry gulch at the bottom of the hill, where Jack Van had flung the hammer at the man whose widow he would later marry. We sat in the truck, looking up the ravine. "Not much to see," Tim said.

"John Olson told me all that's left up here now is stories."

Tim Clawson looked past me up the coulee, where the cottonwoods were shining blue-green in the sunshine. "John's right," he said, and started his truck and drove me back up the rise to my car.

The Bullrider and the Outlaws

They lay in silence, with bitterness between them. Then Elsa said, "Fourteen years ago you were in the liquor business and you got out. Now you're back in it worse."

— Wallace Stegner, *The Big Rock Candy Mountain*

Ahead at the tiny Canadian customs station at Regway, just north of Outlaw Coulee, there was some sort of ruckus. A young officer was arguing with a big Native American man who had gotten out of a dusty red Cadillac with Alberta plates. SUPPORT YOUR LOCAL RODEO a bumper sticker on the Caddy said. A sticker beside it announced, BULLRIDING IS NOT CRUEL TO THE BULL — BUT THE RIDER MAY WIND UP WITH A BROKEN NECK. I rolled down my window to eavesdrop on the dispute.

"So you purchased this in the States?" the customs officer was saying, peering into a package about the size of a shoebox.

"Down at Little Big Horn," said the owner of the Cadillac, who looked to be about my age. "I was visiting my brother and he took me over to tour the battlefield. I picked this up at a souvenir shop there."

The customs agent reached into the box and took out a knife about a foot long. He shook his head. "This is prohibited because of the blade length," he said. "If you'd like to see the regulations, I'll show them to you."

"Oh, I believe you," the driver said. "It's supposed to be a replica of the long knives Sitting Bull and his men carried into battle with them. I thought my son would get a kick out of it. You know, put it on his wall at college. It cost me thirty-five dollars. What should I do with it?"

"Well, you can take it back where you bought it and see if you can get your money back."

The driver laughed a big, jovial laugh.

"I have to be back at work in Calgary tomorrow. I don't have time to turn around and go clear down to Little Big Horn again. That's a good day's drive."

"You could leave it here with us and we'll hold it for you until you come south again.

"NAFTA doesn't cover it?" the man from Alberta said, chuckling at the notion. "Or the Jay Treaty?" He was having some fun.

"I'm afraid not," the officer said.

"Okay, I'll leave it and pick it up the next time through."

They went inside, and the owner of the Cadillac filled out a form and left the replica of Sitting Bull's long knife with the customs officer. Then he drove off north while the officer checked me through. "I told him if he was interested in Sitting Bull and had a camera, to stop a mile up the road, where the first coulee cuts across, and photograph it," the officer said. "That's where Sitting Bull camped when he and his army retreated up here to the Canadian breaks and hid out after Little Big Horn."

Sure enough, where the dry streambed ran under the highway, I found the Cadillac sitting in a pulloff. The driver was leaning against the hood, eating an apple and looking back up the winding gulch the way Tim Clawson and I had looked up Outlaw Coulee a short while ago. I eased in behind the Caddy and went up and introduced myself to the big man, who shook hands genially. He told me that his name was Fred White Killer and that his great-grandfather had fought with Sitting Bull. "Don't worry, though," Fred said with that same big laugh. "So far as I'm aware, none of the White Killers have killed anybody since we polished off old Custer." He reached into a paper bag on the front seat and got out another apple and tossed it to me.

While we ate our apples Fred told me that for thirty years he'd followed the rodeo circuit in the northern states and southern Canada as a bullrider. A few years before he had retired after breaking his forty-fifth bone on his forty-fifth birthday, which he'd regarded as an omen. Now he owned a horse ranch near Calgary. He said to me, "The knife really was a present for my son. He goes to college in Vancouver, and I thought he'd get a kick out of it because of our family name. That's the name I rode bulls under on the circuit. Fred White Killer."

I laughed and so did Fred, who was at least six feet six and weighed about two hundred and fifty pounds. He wore jeans, snakeskin cowboy boots, a blue denim shirt with red and yellow embroidery on the cuffs and collar, and a beaded buckskin vest. He looked as though he could still ride a bull if the occasion called for it.

"So that's where Sitting Bull and your great-grandfather hid out. Up in that draw somewhere."

"So I'm told," Fred said. "I'm glad it was them and not me. Can you imagine scouting around in this country for something to eat? For fresh water, even? Not to mention the winters. Of course the Canadians kicked Sitting Bull and a couple of thousand other Sioux out of here a few years later anyway, made them all go back to Montana and surrender. My son's a history major and could tell you all about it. That's another reason I wanted that damn knife for him. So he could tell his college buddies it was Sitting Bull's, and his old man, the rodeo renegade, got it for him at Little Big Horn."

"Do you think Sitting Bull would have declared that knife back at the border?"

"Not by a long shot he wouldn't have. But look what happened to Sitting Bull. He wound up a curiosity in Buffalo Bill's Wild West Show. Sort of like me, come to think of it."

Fred laughed again, but I sensed an underlying seriousness in his words.

"So you wouldn't recommend the rodeo circuit to an ambitious young man starting out these days?"

Fred shook his head. "Nope. The circuit was good to me" — motioning to the Cadillac — "it's what's paying for my son's education, and my daughter's, too. But let's put it this way. Fred Junior grew up around horses and could ride with the best by the time he was seventeen, the age I was when I joined the circuit. But the day he told me he wanted to go to college was a happy day in my life. Those forty-five bones I broke? It was just pure luck that one of them wasn't my neck. I landed on my head enough, I can tell you that. My kids tell me that's where I got my weird sense of humor from. They're wrong about that, though."

"They are?"

"Yes, sir. Sense of humor is something I've always had, which was a real good thing. Being Indian and trying to break into the circuit back when I started out, in the early sixties? A lot of times, a sense of humor's all that kept me going, partner. Back in those days, the broken bones were the least of my worries."

This country was as rough and broken a landscape as any I'd ever seen, with blind draws and false gaps leading nowhere, looming sandbanks and clay banks honeycombed with caves, impossibly jumbled hills with sage running thick up their sides, arid streambeds that could become deathtraps in scant minutes during flash floods, and abrupt upheavals of stone worked into fantastical geometrical shapes by the western wind and snow and rain.

At Big Beaver, where Tim Clawson once worked at Poi Wing's general store, I turned south, following their old turkey-smuggling route through more breaks and coulee mazes to the border and Whitetail, which consisted entirely of a small wooden U.S. customs building. Here I met Agent Rick Gilmore, a born enthusiast with a boundless interest in anything having to do with the Montana breaks and local outlaws. His three-room customs cabin, open from 9 A.M. to 9 P.M., reminded me of a combination library, local museum, and hunting camp. Everywhere were books, old photos, deer and antelope head mounts, assorted memorabilia from earlier times. A copy of A. B. Guthrie's autobiography, *The Blue Hen's Chick*, lay open on a desk with the hot noon sun pouring on its pages. Rick was just fixing lunch, ravioli and sandwiches, and we ate together at a small table in the sunshine.

"This seems like a good place to get some serious thinking done," I said.

"It is," Rick agreed. "There aren't many distractions. Yesterday ten antelope ran by on that little hill just outside the window while I was eating lunch, and not ten cars crossed into

Montana here the entire rest of the day. Whitetail is one of the most remote entry ports in the U.S. That's what made it such ideal rustling country. The Outlaw Trail up from Mexico, through Robbers' Roost and Hole in the Wall, ended right here in the breaks."

"How did the outlaws get through the winters?"

"They lived in caves enlarged from wolf dens dug into the coulee banks. A posse could ride by twenty feet away and never suspect they were there."

"The country just north of here still has the feel of a frontier. I didn't count five buildings between here and Big Beaver."

"It's a very young territory. I'm only the fourth U.S. customs inspector to be stationed in Whitetail. The first, a man named Noland Armstrong, came to a bad end. Armstrong was a mounted inspector, and a well-liked man in these parts. But one January night in 1900, he got drunk at a local saloon and wandered into the street and began shooting at the moon. A trigger-happy local deputy sheriff cut him down on the spot with a shotgun."

"What happened to the deputy?"

"Well, he narrowly escaped being lynched. Then four months later he had a falling-out with a neighbor, threatened the investigating sheriff, went for his gun, and was shot dead himself."

"It sounds as though half the people who settled up here were outlaws."

"A good many of them were," Rick said, "and of course I'm fascinated by them. What really interests me, though, is preserving the history of this place."

Rick Gilmore looked around at the prairie and the breaks to the north. "You know who I admire most? Not the outlaws, not the lawmen, not the ranchers or homesteaders. The people I admire most are writers like A. B. Guthrie and Wallace Stegner and Norman Maclean, who've preserved some of that past for us in their books."

West of Glasgow I caught my first glimpse of the foothills of the Rocky Mountains, and they caused me a pang of loneliness for

Phillis, with whom I had seen the Rockies in the summer of that abbreviated venture west to graduate school in California. I was as exhilarated by the sight of the tall western mountains (far higher and more massive than I'd remembered from my college hitchhiking trip) now as then. And I was glad to learn from a roadside historical marker in Malta that near here in 1901, Kid Curry, a member of Dutch Henry's Wild Bunch, boarded the Great Northern Flyer when it stopped for water, unceremoniously blew the door off the express car, stuffed $40,000 into his saddlebags, splashed across a nearby ford on the Milk River — and was never heard from again.

The fabled Great Northern has now been swallowed up by the Burlington Northern, whose ore-loading yard I'd visited back in Superior, Wisconsin. And when about ten that evening I pulled into Havre — pronounced *Have*-er, accent on the first syllable — the headquarters of the BN from Minot, North Dakota, to Whitefish, Montana, I felt again, as I had at Seney and Marquette and Ely and Plentywood and a score of other North Country frontier towns, that somehow I'd come home. For it was here in Havre that one of my all-time favorite fictional characters, the tragically flawed Bo Mason of Wallace Stegner's great novel *The Big Rock Candy Mountain*, sold the whiskey he smuggled down over the Saskatchewan border, at a time when Havre was not only a rail center on the Great Northern but, very possibly, the wildest town in Montana.

I checked into the first motel with a vacancy on the east side of town and promptly embarked on a self-guided nocturnal walking tour, wending my way through the Burlington Northern rail yard along the Milk River and on down the entire length of Main Street, past a couple of dozen motels, fast-food joints, and gas stations — where once a solid row of speakeasies beckoned to the wayward and respectable alike. I grabbed a hamburger at an all-night truck stop, then struck out for Pleasure Hill, where Bo Mason was known to hold court with a cigar in one hand and a full glass of his own bootleg whiskey in the other.

The term "red-light district" is said to derive from the glowing red signal lanterns left on the Pleasure Hill whorehouse stoops by Great Northern trainmen. Before phones were in common use in Havre, call boys, as they were dubbed locally, were hired to race up the hill and fetch back reveling train crews to work on short notice. In those days this was a lively spot, with a complex network of secret tunnels leading down to Havre's main-street speakeasies. Tonight, though, the Hill was just another peaceful, darkened neighborhood of comfortable one-family houses. Except for a couple of mostly deserted downtown bars showing late-night baseball games from the West Coast, I didn't find a vestige of Havre's wide-open nightlife this evening.

The Ghost of Wallace Stegner

Let it be, at least for a good long while, a seedbed, as good a place to be a boy and as unsatisfying a place to be a man as one could well imagine.

— Wallace Stegner, *Wolf Willow*

I was dog tired that night and slept until seven o'clock the next morning, when I came awake hard to the rumbling of a train going by on the Burlington Northern. The motel coffee didn't perk me up much, and neither did the two additional cups I had at the café next door, where I ate breakfast. Time, I thought, to slow down. For starters I'd hunt up the border ranch where Wallace Stegner spent his boyhood summers with his gentle, nurturing mother and his erratic father, for whom the great American dream eventually went as bad as the rotgut whiskey that he, like Bo Mason, smuggled over the line to Havre.

The American customs official on duty at Willow Creek, north of Havre, hadn't heard of Wallace Stegner, but the Canadian officer said that I should go to Eastend, Saskatchewan, sixty miles to the northeast, and inquire there. My itinerary had impelled me westward or northwestward for the past month, and I found myself reluctant to backtrack, even though I would be traveling in Canada and through new country. But I was excited, too, not merely by the anticipation of another literary quest into new country but by ties of friendship as well.

Heading across the high, treeless prairie north of the border, I thought about my first personal contact with Wallace Stegner. For years the novelist had summered in northern Vermont not far from my home, and when he dropped by to visit one morning, I mistook the tall stranger at our door for the town's new tax assessor and greeted him with a distinct coolness. We became good friends, however, staying in touch until Stegner's death at eighty-four in May 1993.

Eastend lay in a wide loop of the Frenchman River, and the hills rising sharply above it were ribbed all the way to the top with exposed strata of kaolin clay, gleaming white as snow in the late-morning sunshine. My first order of business was to hunt up the shrubs for which Stegner named his memoir, *Wolf Willow*. At certain times of the year the odor of these bushes permeated the entire community — "sticky-smelling," Stegner called it — and it was this resinous scent that told him he was home again after being away for thirty years. Today the banks of the Frenchman were still lined with willows, but the gray-green leaves were powdered with whitish dust, and I had to rub several between my thumb and forefinger to catch a hint of that distinctive redolence.

The town itself bore little resemblance to the assortment of frontier shanties where Stegner grew up. The streets were neat and prosperous-looking, with a new Wolf Willow Health Center, a White Mud Grain Co-op, and a modern bank, where I was given directions to Pete and Sharon Butala's ranch a dozen miles south of town. Sharon, I was told, was a novelist who'd written extensively about the Saskatchewan-Montana border country herself and had helped establish Eastend's local arts council, which purchased the house in town built by Wallace Stegner's father and sponsored several writers-in-residence there. A teller at the bank drew me a map to the Butala place, and I headed out of town on a network of dusty gravel roads, past brimful irrigation ditches tapping the Frenchman River, into a region of rolling, treeless hills that all looked alike. Soon, I was thoroughly lost. I stopped to ask directions at four different ranches, but not a soul was home at any of them. Every member of every rural household in southern Saskatchewan seemed to be out harvesting faraway fields today. The silent, deserted yards and houses reminded me of a long-ago *Twilight Zone* episode in which a whole townful of people was swooped up by aliens one fine summer morning. I branched out onto several ever-dwindling side roads, and around noon, by sheer luck, I spotted the Butalas' mailbox. Sharon and Pete, who had just come in from the fields for dinner, pulled an extra chair up to the table as if they'd been expecting me for a week.

While we ate, Sharon told me that she'd never met Wallace Stegner. They had corresponded, and more than once she'd invited him to revisit Eastend and see the renovated family home. This past spring he'd agreed at last to make the trip during the summer. Then in May he had been in a fatal automobile accident; the Butalas, particularly Sharon, had been devastated.

Although the meal and talk had seemed unhurried, when one o'clock came it was time to return to the fields. Pete gave me directions to the vicinity of the old Stegner ranch and reminded me to keep my gas tank full, and Sharon walked me out to the car.

"Wallace Stegner would have enjoyed meeting you folks," I said. "I'm sorry he never made it back here."

Sharon hesitated. Then she said quietly, "Actually, he did. Do you believe in ghosts?"

I was mildly startled. "Well, I've never had an encounter with one."

"I have," Sharon said. She handed me a manila envelope. "Read this when you have time. You'll see what I mean."

I am not entirely certain that I found the old Stegner ranch, or even the U.S. border. If not, though, I came very close. I parked beside a faint track in the prairie that I couldn't follow any farther without a four-wheel-drive vehicle and sat on the hood in the relentless sunshine and felt as though I was a million miles from anyone. There wasn't a house or a shed or a fence or a tree in sight, and the only moving object besides the yellow-green shortgrass bending in the constant wind was an antelope on a distant hilltop, its African looking white, buff, brown, and black standing out sharply against the sky. The prairie emptiness that Wallace Stegner once described as "almost frightening" was a nearly palpable presence, like the wind.

In this most fitting of settings, I opened the envelope that

Sharon had given me and read the straightforward, moving tribute she had written to Stegner, which concluded with a mystifying story of her brush with the supernatural on the night he died:

> As I lay there wide awake in the country night's silence, so quiet the hush becomes a sound in itself, an encapsulating, velvety hum, I tried to assimilate the news of his loss. I found I was not really clear how I should feel, since I'd never met him and could hardly call him a friend, and yet a dark hollow was growing inside me as if he had been my intimate, long-time companion, as if I had loved him. I thought ruefully, as one does, of lost chances, of my own failings with regard to him, because I had never gone to California to meet him and now never would. Gradually, I realized that his spirit was there in the room with Pete and me, and this was how I knew so surely he had died. It seemed that if we could not meet in life, he had come for a visit before he left for whatever his new destination was to be. . . . Already his humanity, stretched thin by a long earthly life, was transparent, soon it would rupture and dissolve entirely . . . leaving this gorgeous earth behind and soaring off on the solo voyage from which there'd be no return.

I returned Sharon Butala's essay to its envelope, and by degrees the prairie and wind and grasshoppers returned to me. What to make of such an experience? Bred-in-the-bone skeptic that I am, I could not be sure — except to reflect that Wallace Stegner, himself in many ways the most realistic of men, had nevertheless written in *Wolf Willow* that this was "a country to breed mystical people" — along with outlaws like Bo Mason and his own father.

The Writing Life

It was not until I began the struggle to live and write in my region that I began to be aware of Wallace Stegner as a writer struggling to live and write in *his* region.

— Wendell Berry, "Wallace Stegner and the Great Community," in *What Are People For?*

Along with my Northeast Kingdom friend and neighbor the acclaimed essayist Edward Hoagland, Wallace Stegner had been of inestimable help to me in my career as a writer, and that afternoon, back in Eastend, where I visited the house the Stegner family had lived in during the winters, I thought again of the encouragement he'd given me with my early work. Though he hadn't liked *Disappearances*, my wild and woolly first novel, Stegner had praised my short stories and later novels generously; with his praise, I felt a certain vindication for those long, uncertain years of struggle to get them right.

To back up a few years, after leaving my job in the woods with Jake, I returned to my Kingdom stories with a vengeance, revising them all again — and again and again. It became obvious that the editors at the big magazines simply weren't interested in what I was doing. But one joyous fall afternoon in 1972, a few months after the birth of our son, I got an acceptance from *The Cimarron Review*, a literary journal in Oklahoma. It had been raining for three days and the mountains were shrouded in dense fog, but quite by coincidence, as I headed back from the mailbox, so elated I was scarcely aware of my surroundings, the wind blew the clouds away, and under the strong late-afternoon sun the mountains were a solid bank of red and gold all the way from Mount Mansfield north to Canada. That was a great moment in my life — standing in my beloved border country with the acceptance in my hand and thinking that even if I never got another one, I would always remember this experience.

Soon afterward I began an early draft of a short novel, *Where the Rivers Flow North*, based closely on Jake Blodgett's life and times. The book, which took five years to write, attracted the attention of Don Congdon, a highly respected veteran New York literary agent, who had grown up in a rural community much like my hometown in the Catskills. Don understood what I was trying to do with my North Country fiction. When he circulated *Where the Rivers Flow North*, however, several publishers told him that I had to write and publish a full-length novel to establish an audience. I was disappointed, but Don encouraged me to keep writing.

In one respect, then, I started *Disappearances* in desperation. I certainly wanted to write this story of a Prohibition-era family of French Canadian whiskey smugglers who, instead of dying, literally disappeared — just as most of the Kingdom's hardscrabble hill farmers, horse loggers, moonshiners, and individualistic storytellers, people whose like would not be seen again, were disappearing. But I knew that turning out even a rough first draft would take months of night-and-day work, and I'd already been writing hard for eleven years without a book.

"Do you think I should quit doing social work and just write?" I asked Phillis.

"Why not?" she said immediately.

It wasn't quite that simple. We had to sell our Vermont home overlooking those gorgeous mountains and rent a cramped house in a nearby village. Leaving our first home, where our son and daughter had been born and I'd written my first stories and *Where the Rivers Flow North*, was heart-wrenching. Nor was I eager to quit my job at the social agency. But I was beginning to realize that my North Country muse was a jealous and demanding one, and I had damn well better listen to her — and to Phillis, who was telling me the same thing — if I intended to make a career of writing fiction.

Throughout the fall and winter of 1976, I worked on *Disappearances*. Sometimes I lugged the manuscript up the street to the village library, located smack on the Canadian border, which was represented by a black line running down the middle of the reading-room floor. I deliberately chose a table straddling that painted line.

In the spring of 1977 Don Congdon sold the novel to Viking along with *Where the Rivers Flow North*; we were able to make a down payment on the house we still live in, not far south of the border. So the book I wrote with the proceeds of one house purchased another — a fair enough tradeoff.

Notes from the Milk River Breaks

In the American West, men came before the law, but in Saskatchewan the law was there before settlers, before even cattlemen, and not merely law but law enforcement.

— Wallace Stegner, *Wolf Willow*

2:00 P.M. Once again I have a decision to make. Do I take the bad gravel road to the town of Milk River, Alberta, one hundred and fifty miles to the west through a wilderness the Butalas had told me experienced cowpunchers avoided because of the inhospitability of the breaks? Or do I skin back down to Havre and push west on my old familiar friend, Route 2? I look at the wilderness road on my Rand McNally: like so many of the routes I've traveled since Maine, this isn't a "blue highway" so much as no highway. From Willow Creek to Milk River, it's designated by two close-set, parallel, light gray lines with nothing between them: ═══. *Unpaved*, the map legend says; and that does it. The wilderness road wins hands down.

2:30 P.M. The Milk River and the U.S. border lie just to the south. To the north, plainly visible in the clear western light, are the Cypress Hills, misnamed by a British surveyor who mistook the stunted jack pines scattered over otherwise barren ridges for cypresses. Here in May of 1873 the atrocity known as the Cypress Hills Massacre took place; a band of drunken American whiskey smugglers armed with brand-new Henry repeating rifles slaughtered scores of Assiniboins, burned their village, and rode off well satisfied with their day's work. Canada's nationwide outrage over the massacre resulted in the establishment of the Northwest Mounted Police, who were dispatched the following year with two hundred and seventy-five hand-picked officers, seventy-three freight wagons, one hundred screeching Red River carts, two nine-pound cannons, and scores of cattle to eat along the way. By the time this dashing brigade reached the Cypress Hills, their ranks had been decimated by disease, injuries, and near-starvation. Their cattle were gone, many of their horses had been stolen by rustlers, they'd been half-frozen by late-summer blizzards, lost for days on end in the breaks, and threatened by the Blood, Blackfoot, Piegan, and Sioux. No matter. The Mounties reached the whiskey-trading post known as Fort Whoop-up just before winter and immediately set about suppressing the smuggling, negotiating an enduring peace with the Indians, and arresting the local rustlers and horse thieves — all with remarkably little fanfare or violence. They'd been given a job to do and they did it. In Canada the Outlaw Trail, which informed so much of the Montana territory just to the south with its renegade reputation, was a short-lived phenomenon.

3:00 P.M. Crossing into Alberta and Mountain Time, I slow down to read a road sign: ATTENTION. STOCK AT LARGE. The sign enhances my notion that in entering the Rocky Mountain time zone I've gone back to an earlier frontier era; yet after driving another twenty miles through terrain as empty as it was in the early fall of 1874, when that fever-ridden, diminished band of Mounties rode through it, I'm oddly relieved to meet a dozen Hereford cattle in the road.

3:30 P.M. I like the intense heat of this country and find myself using the air conditioner only sparingly, but I regret not buying a couple of Cokes back in Eastend. Earlier today three different people admonished me to keep my gas tank full. No one told me to lay in a stock of soda, and I'm not about to risk drinking out of the alkaline-looking roadside streams.

4:00 P.M. TEXAS GATE AHEAD, the sign says. A barbed wire fence runs down tight to the road on each side, but I don't see a gate of any kind, just a grate consisting of several spaced metal pipes laid at right angles to the road and flush with its gravel surface, which the range cattle evidently won't or can't cross. Soon afterward I come to the shells of half a dozen abandoned houses and a windowless one-room school. The tiny ghost hamlet reminds me that as John Olson back in Plentywood said, the overwhelming majority of homesteaders, ranchers, trappers, outlaws, farmers, teachers, doctors, ministers, and entrepreneurs who came to this harsh country along the Montana-Canada border within the last century sooner or later departed.

4:30 P.M. Over the Sage River, whose sandy bed contains far more sage than water. Through Onefour, designated by Rand McNally as a town but consisting of nothing more than a few scattered ranches. At the junction of Alberta 501 and 502, both gravel lanes, I'm tempted to swing north a few miles to Manyberries for a cold drink and the satisfaction of visiting a town of that distinguished name. But I'm not going to die of thirst, as some of those early Mounties nearly did, and in the end I skirt Manyberries to stay close to the Milk River.

5:00 P.M. On I drive into the crystalline afternoon, and on and on and on, basking in the sense of well-being that comes over me in wild country yet aware, too, of the tenuous nature of that state of mind, which in this drought-dry, untraveled region could, with two flat tires or an overheated radiator, change instantly to distress. Sometime around five-thirty, I'm relieved to see a small sign with an arrow pointing south: WRITING-ON-STONE PROVINCIAL PARK. MILK RIVER HOODOOS.

5:30 P.M. I follow a footpath toward some rapids, emerging into the strangest North Country topography I've seen yet. Over the eons the river has worked its way down through layer upon layer of red, yellow, and brown sandstone. The bank is crowded with hundreds of mushroom-shaped columns of multicolored sandstone, ten to sixty feet high and two to twenty feet in diameter, looking almost man-made in their uniformity. Their neat stone berets weigh tons apiece. "Hoodoo" is the perfect name for these configurations, each one a geological conundrum formed by just the right combination of water and wind twisting through the sandstone canyon. The word itself, African in origin, connotes a jinx or bad luck, which, no doubt, is exactly what the settlers of this area had every good reason to fear when they first set eyes on the hoodoos. To me, though, wandering through this forest of gigantic stone mushrooms glowing in the low rays of the sun like the columns of some strange lost city, the hoodoos are more august than ghostly, and I'm happy to have come here at the end of this good day in the breaks along the Outlaw Trail on the Montana-Alberta border.

A Honeymoon in Shelby

There was a commotion that got loud and moved outside to the windy street. Two men [from Shelby] . . . came at each other, locked outstretched arms and pushed, circling slowly as if turned by the prairie wind. They tired and revolved slower, but neither let go or fell down. A police car drove up and honked. The fighters went to the squadcar, both leaning on the window to listen. After a while, they slumped off in opposite directions, and that was the end of it.

— William Least Heat-Moon, *Blue Highways*

Somehow, coming across the wonderfully weird Milk River hoodoos inspired me to drive straight south on a bad back road, not shown on my map, with the idea of sneaking back across the border into northern Montana's Sweetgrass territory. At sunset I crossed the Milk on a high, exceedingly narrow iron bridge, pushed on through more breaks and prairie, and dead-ended in the barnyard of an abandoned homestead. I backtracked, tried two more roads that petered out into cattle paths, and conceded that I'd have to go back to Writing-on-Rocks Park and take the main road south from Milk River. At the Sweetgrass port of entry, I stopped to visit with the American customs people, who seemed eager to hear all about my trip. After my long, lonely drive through the breaks, I was ready to visit with anyone who had the time and didn't leave the customs station until nine o'clock.

I wondered about motel accommodations, but I wasn't overly concerned; on this kind of jaunt you can't spend time fretting about where you're going to stay. I hadn't had to use the back seat of the car yet, but before leaving home I'd tossed an old army blanket into the trunk just in case. I decided to push on to Shelby, on Montana's so-called Hi-Line, or border country, and there I located a down-at-the-heels motor court directly across from the Burlington Northern tracks. In the parking space next to mine sat an eighteen-wheeler with the words Douglas Explorations emblazoned on the cab and a complete oil-drilling rig chained onto its trailer. The space on the other side of my car was occupied by a spring-shot maroon Pontiac showing lots of hard mileage and rust. Tattered shreds of colored crepe paper hung limply from the Pontiac's door handles, and a hand-lettered JUST MARRIED sign clung to its crumpled rear bumper. The lights were on in the honeymoon suite, Number 6, from which loud music was issuing forth.

My own room, Number 5, had no lamp over or beside the bed — whoever was reading in the North Country these days wasn't doing much of it in economy motel rooms — so after a quick burger at the truck stop next door, I clicked out the overhead light and climbed into bed. As if on signal, a clangorous hammering started up in the BN freight yard just across the highway. The ringing vibrations shook the flimsy walls of my room, as did the bass from the jacked-up boom box in the adjacent honeymoon boudoir. After a prolonged interval I managed to drift into a restless half-sleep, which was soon disturbed by a jangling telephone. Real or imagined, the ringing phone set off a nightmare that I recognized as a nightmare but couldn't shake myself out of, in which Phillis was trying to get through to me about our kids.

What jolted me fully awake was a shout from next door. Apparently the honeymoon had hit some turbulence; angry screams were now emerging from Number 6. Meanwhile, the trucker in Number 4 was pounding on the thin beaverboard partition eight inches from my head and hollering for quiet. I sat bolt upright, reached for the bedside lamp, remembered that there wasn't any, stumbled across the floor, and snapped on the dim overhead light.

It could be worse, I reflected. I could, for instance, be in the brig at the command base in Minot as a result of my run-in with the United States Air Force on the unmarked missile site. I pulled the single wooden chair in the room out under the ceiling light and repaired to my stock of traveling books. Through with *Wolf Willow* and Stegner territory, I cracked open *A River Runs Through It*. I had just gotten caught up in Norman Maclean's stately cadences when all hell broke loose in Shelby, Montana.

From Number 6 the music and shouting rose several decibels. The clanging from the BN yard was now accompanied by deep thudding collisions. I had an irrational and nearly overpowering urge to call Phillis to make sure the kids were okay, but a phone ringing suddenly in the middle of the night is not a reassuring sound to a woman with children away at college, elderly parents, and a footloose husband on the road, so I made myself put the receiver back on the hook. As I did, a long shriek accompanied by a string of unhoneymoonlike expletives arose from Number 6. In Number 4, Douglas Explorations was pounding steadily on the adjoining wall, bidding fair to put his fist right through into my room. "What the hell is going on in there?" he yelled.

"Nothing's going on in here," I yelled back. "I'm reading *A River Runs Through It* by Norman Maclean."

"Shut up! Shut up!" a young man's voice roared from Number 6, as though Douglas Explorations and I were the ones keeping everyone awake.

"I'm calling the police," Number 4 shouted. "I have to get to Wyoming in the morning."

"Fuck Wyoming!" 6 shouted, to the accompaniment of wild female laughter.

Hoping to take advantage of this temporary marital rapprochement to get some sleep, I again shut out the overhead light and returned to my bed. This time as I started to drop off, a freight train seemed to be barreling through Number 6. It didn't entirely wake me up, but the pounding on my door a few minutes later did. I threw off my single cigarette-scorched blanket and ran to the door, sure that it was Douglas Explorations. Instead, two uniformed patrolmen stood there. "Wrong room," I said. "Commotion's next door."

The younger of the two officers grinned. "Oh, that. I guess the marriage is off to a rocky start. We've already told them to pipe down. Is this your car here with the Vermont plates?"

Now what? My first fleeting thought was that the air force had run me down at last, decided to press charges, and teach me a lesson. Or, God forbid, was there really trouble at home this time — my night-terror, the imagined phone call? But no. The police had just stopped by to tell me that my trunk lid had popped open and they thought I'd want to know.

I thanked them, told them that above everything I didn't want to lose my fly rod — heaven knew what the honeymooners would do with it if it fell into their hands — and got into my hunting jacket to go shut the lid.

"If you're going to use that pole yourself you'd better start soon," the older cop said. "Snow's predicted in the foothills tomorrow night, and that'll be the end of your fly fishing."

They circled the motel parking lot once in their patrol car, and the instant they pulled back out onto Route 2 the shouting and name-calling and wild music started up again. I flopped back down on top of the holey blanket, as exhausted as I could ever recall being. The stereo next door thrummed through my bed frame. Shouts and pealing screams split the night. Telephones rang steadily. Over in the Burlington Northern yard, a deafening explosion seemed to go off — for all I knew, a Minuteman missile. Soon a locomotive started up directly outside my door. My whole room shook with its rumbling reverberations, and when I got up and lurched across the shuddering floor to the window, the sky was lightening and smoke was pouring out of the twin stacks of the Douglas Explorations eighteen-wheeler.

I stood there barefoot, looking out the window into the dawn. For the first time since leaving Vermont, I was completely played out. I felt desperately homesick. It crossed my mind to throw in the towel and cut for Vermont, leave the last leg of my trip for the coming spring, when I'd be rested and everything would look brighter. Just realizing that I had that option, however, was liberating; by the time I'd taken a shower, shaved, and packed up my books and maps, I felt downright emancipated, not to abandon my trip through the North Country but to forge on ahead and see what the Rocky Mountains, looming just ahead to the west, held in store.

Jimmy Black Elk

Where all this took place was in that remote part of Montana near the Canada border and west of the Sweetgrass Hills. That is called the Hi-line, there, and it is an empty, lonely place if you are not a wheat farmer.

— Richard Ford, *Rock Springs*

"Sometimes survival is the only blessing that the terrifying angel of the Plains bestows," the poet and essayist Kathleen Norris writes in *Dakota: A Spiritual Geography*. Most of the plains dwellers I'd met since crossing the Red River would readily agree with her. For that matter, so would most North Country natives. From Maine to Montana, to survive in America's harsh northern climes you must be resilient enough to fail and fail and fail — with your crops, with your business, with your dreams, with whatever — and still somehow pick up the pieces and keep going.

By the time he reached the broken country between the present-day town of Shelby and the foot of the Rockies, Captain Meriwether Lewis had already learned this lesson one hundred times over. Lewis had temporarily split off from his partner, Captain William Clark, to trace the Marais and Cut Bank rivers up to their sources, hoping to discover that they rose north of the forty-ninth parallel, thus extending the northern perimeter of the American territory bought in the Louisiana Purchase. They didn't, though. Not many miles from Glacier International Park is the site of Camp Disappointment, where Lewis and his scouts turned south again to join the rest of the expedition, hastened on their way by fear of an encounter with the Blackfeet, whose nearby presence they strongly suspected from the skittishness of the local game. Disappointed Lewis couldn't help being, but once again he and Clark tightened their belts and pushed on west toward the Pacific — as I, in my middle-aged writerly way, now did myself.

The Blackfeet are past masters when it comes to survival. After the Battle of Little Bighorn, at which Blackfeet warriors distinguished themselves for their particular bravery, followed by the arrival of the railroads, bringing settlers in large numbers, they withdrew to northwestern Montana and became guides, cowboys, ranchers and, in many instances, artists and craftsmen. At Browning, the Museum of the Plains Indians showcases some of the finest craftwork in North America, by the Blackfeet, Crow, Northern Cheyenne, Sioux, Assiniboin, Arapaho, Shoshone, Nez Perce, Flathead, Chippewa, and Cree. What especially caught my eye was the magnificent beadwork, colorful as a Rocky Mountain wildflower meadow, on a display of turn-of-the-century moccasins, deerskin vests, and jackets.

A Native American arrow maker at the museum who was chipping out flint points as lethal-looking as any that had Custer's name on them one hundred and twenty years ago said to me reflectively. "You know, that was all so very recent, historically speaking. The bat of an eye, really. It's scarcely any wonder that some of our people have had difficulty adjusting to what amounts to a whole new culture in what, just four or five generations."

A few miles north of the museum, I picked up a disabled hitchhiker, a young man on crutches who'd lost his left leg from the knee down. He began talking as soon as he was in the car, telling me that he was on his way home from his girlfriend's in town, that she'd kicked him out just this morning. He was carrying a paper bag with some clothes and personal effects and a twelve-pack of Coors beer, which he broke open with gusto, keeping one beer for himself and handing one to me. His name was Jimmy Black Elk, but everyone except his girlfriend called him Pegleg since his "run-in with a chainsaw" two years before. He was nineteen, had dropped out of school at fifteen to work in the woods with his father and brothers, and now had a small monthly disability payment. He'd tried going back to school, but it hadn't worked. The local sheriff — a Blackfoot named Rob Powell — had hooked Jimmy up with a correspondence course to study for his high school equivalency diploma during one of two stays in jail — once for resisting arrest by a game warden, once for drunk and disorderly — but he hadn't followed through on the program.

After Jimmy's release, Sheriff Powell had urged him to try his hand at craftwork. He showed me the beaded headband he was wearing and a foot-high elk he'd carved from a block of ponderosa pine and given to his girlfriend, who'd made him take it back that morning. "Can't hunt 'em anymore so I might as well carve 'em," he said with a certain grim satisfaction.

In the past Jimmy Black Elk had done a little guiding, but since his chainsaw accident "nobody wants a pegleg for a guide." These days, he admitted, he was drinking a lot, watching the "afternoon stories" on television, and "feeling good and goddamn sorry for myself." He was bitter and angry, yet he had a certain cheery insight into his dilemma, and into himself, too. "I've got an artificial leg at home," he told me, cracking open another beer. "'Prosthetic device,' they call it. I don't wear it much. My girlfriend gets on my case, says the reason I don't is I want people to pity me, give me rides and buy me beer and such. Fact is, that goddamn thing twangs my stump every step I take. *You* cut off your leg with a chainsaw and try strapping on a 'prosthetic device' that twangs your leg. Here. Have a cold one."

Jimmy thrust another beer at me and asked what I did for a living, besides ride the roads. When I told him, he said he'd go buy one of my books except that he never did learn how to read very well, though his girlfriend read all the time. He seemed to find that amusing, too.

"Next place on the right," he said.

Ahead, off the road a few hundred feet, sat a shack with an abandoned school bus in front of it, resting on tireless wheel rims. Jimmy told me not to bother to turn in but didn't seem unhappy when I did. "I live in the bus," he said. "Want to take a gander at it?"

The bus had a stovepipe sticking out of its roof and newspapers taped over some of the windows for curtains. The back window was missing entirely, and over the space Jimmy had fitted a rectangular section of cardboard. He'd put up some centerfold pinups and some pictures from outdoor magazines. There was an unmade cot with a tattered quilt bunched up on it, a long counter crowded with empty tin cans, a Coleman lantern, and a lever-action hunting rifle. He told me he'd taped the cardboard over the rear window after shooting it out with his thirty-thirty when a mule deer appeared on the knoll across the road, but then he'd gone and missed the deer.

"Story of my life," he said, opening another beer.

At the front of the bus sat a workbench where he carved his wooden animals. He had a few "seconds" on hand, as he called projects that for one reason or another hadn't turned out satisfactorily, including one of a leaping trout, which I asked if I could buy.

Jimmy Black Elk shook his head. "I never sell seconds. Use 'em for kindling. But you can have that fish if you want it. Carving's just something I do to pass time, anyway. I don't work at it that much."

I looked at him incredulously. To the degree that I could judge, Jimmy's animal carvings were first-rate. But he seemed to be in a state of despair or close to it, already well into his third beer since I'd picked him up, and back on the subject of his girlfriend. How would she like to gimp around on one leg, he demanded.

"Not to press the point," I said, "but how much would that trout sell for if it weren't a second? If it suited you?"

He gave the carving a bleak look. "About twenty bucks."

"Can I ask you one more question?"

"Shoot."

"If you could do anything at all for a job, what would it be?"

"That's easy. I'd fish and hunt and trap for a living. Live off the land, same as my ancestors did. Course, that's impossible in this day and age. Specially for a pegleg."

"What would the next-best work be?"

Jimmy thought for a minute. Then he gave me a beery grin. "There isn't any next-best work, to my way of thinking."

"Not carving your animals?"

A cunning look came across Jimmy's face. "Let's say somebody came along and told you you couldn't write no more, but you could do the next-best thing. What would you do?"

"I can't imagine."

"Neither can I," Jimmy said, shoving the carved trout into my hands. "Don't try to leave any money for this. It's a present."

On Setbacks and Windfalls

VERMONT WRITER SHOULD DISAPPEAR

— Headline over a review of *Disappearances* in the *Montreal Gazette*, 1977

My visit to Jimmy Black Elk's got me thinking about some of my own early setbacks. Although I never had a misfortune in any way comparable to Jimmy's loss of his leg, I quickly discovered that it's one thing to write and publish two books and something else again to settle in for the long haul as a novelist.

Almost as vividly as I remember the glorious fall afternoon of my first story acceptance, I recall the fall morning in 1977 when I picked up the *New York Times Book Review* at the post office and read a sneering and dismissive notice of *Disappearances*, by a writer whose name would be emblazoned in my mind for years to come.

The following Saturday afternoon, just over the border in Quebec, I stopped at a general store and idly picked up that day's *Montreal Gazette*. Staring back at me from the front page of the weekend entertainment section was a reproduction of the jacket illustration of *Disappearances* — under the blaring headline VERMONT WRITER SHOULD DISAPPEAR. The accompanying review, or that part of it I could bring myself to read, made the scathing *Times* notice sound like a glowing accolade.

And in the fall of 1978, after publication of *Where the Rivers Flow North*, the *Times* went off the newsstands for six weeks because of a printers' strike; the interview-review planned for the book was stillborn. Like Jimmy Black Elk, I was beginning to wonder what in hell I'd have to do to catch a break.

Still, at thirty-six I was learning that how we take our bad luck can be a major determining factor in how we make our good luck, and I made up my mind that I wasn't about to disappear. Taking my uncle's long-ago advice to write about the decline and fall of Chichester, I moved the little woodworking village to northern Vermont and made it the setting of *Marie Blythe*, my third novel. Some windfalls came my way, including a Guggenheim fellowship; a New England Book Award for *A Stranger in the Kingdom*; and the wonderful 1994 feature film of *Where the Rivers Flow North*, directed by my close friend, the brilliant independent Vermont filmmaker Jay Craven.

As I continued to develop the geography of my fictional territory, a number of reviewers began to apply the tag "regional" to my work. To me the term "regional" sometimes seemed patronizing, a cheapening of my fiction. By degrees, though, the classification ceased to bother me. If to some critics my Kingdom County seemed antiquated, even quaint, so be it.

The Blackfoot, the Maverick, and the Mountie

What the U.S. Army could not accomplish — the destruction of tribal organization — whiskey traders did with help from Christian missionaries who suppressed the old rituals. The white settlers, moving in after tribal disintegration opened this land, should have erected a monument to the whiskey bottle. The Blackfoot, for example, once hunted an area about twice the size of Montana; now their reservation of steel crosses and Whoopie Burgers doesn't occupy even all of Glacier County. It isn't that Indians lost their land because of whiskey — that stuff they called the Great Father's Milk — they just lost it faster because of whiskey.

— William Least Heat-Moon, *Blue Highways*

I had been awed by the snow-capped peaks of Glacier Park when Phillis and I camped there on our way west to California in 1969, but I had little desire to revisit the park. After all, Niagara Falls had been penultimate North Country at one time, too, and look at that place now, emasculated by grotesque amusement parks, with the river reduced to less than half of its original flow by hydroelectric projects. As soon as I could, I swung off the main route onto a parallel side road, away from the silver Airstreams and Winnebagos, and poked my way up through the foothills along a meadow stream.

The willows and aspens beside the brook were turning yellow, and here and there magpies were perched in them, balancing themselves with their long, flashy tails. I crossed a tributary of the Milk River, tinged with chalky white glacial runoff. Snow from last night's squalls reached far down the sides of the high peaks to the west, and the lichens on the bare rocks just below the snow line had turned a bright magenta; whole cliffsides were lighted up in the sunshine like Vermont hills in foliage season. I came back onto the main highway just south of the park entrance at St. Mary, climbed some more, then started down a steep mountainside toward the border, past several runaway truck ramps angling off the highway and packed deep with loose gravel to stop out-of-control logging trucks that had lost their air brakes. I tested my own brakes, found them tight, and continued down the grade toward the border, knowing that if I needed to, I could stop in a hurry in a region of the country where my life might damn well depend on it.

Rob Powell's ranch abuts the Canadian border on the north and Glacier International Park on the west. When I arrived, the Blackfoot rancher and ex-sheriff of Glacier County was out in a roadside field picking up bales of hay with his farm truck. A tall man in his sixties, rangy as a college center fielder, with short gray hair and a kind, intelligent face, he waved to me as I walked across the field between the bales. He wore jeans, work shoes, a blue shirt, and tough work gloves, which he removed to shake hands with me. He was carrying a wooden-handled hay hook a foot long to heave the sixty-pound bales up into the back of the truck. When I told him that Jimmy Black Elk had spoken highly of him, Rob nodded, smiled, and laid his hook and gloves on the bed of his truck. The Black Elk family down in Browning had been good neighbors to his family for generations, Rob said. (It never failed to surprise me out west, how far away your "neighbors" might live — twenty, sixty, even one hundred miles.)

"My great-grandfather was one of the first local ranchers," Rob told me. "He came up here right after Little Big Horn, and we've been here ever since. It's a small ranch by local standards, only a couple of thousand acres. But since I retired from sheriffing, I've kept busy here breeding beef cattle and shipping the calves for veal."

I asked Rob Powell how living on this section of America's northern frontier had shaped the course of his life and work, and he smiled. "Well, for myself and for many other Blackfeet, living up here in these mountains has given us a special, almost mystical feel for the land and everything that lives and grows on it. But some Blackfeet have been cut off from the land and their past. Frankly, many don't want any different way of life, and the results can be tragic. That's what I found myself dealing with time and again as sheriff. People who would have been perfectly successful in our earlier culture are lost souls in today's. What this means is, police officers in a place like Montana's Glacier County have to be trained to do a lot of listening. Usually, if you're willing to listen long enough, most people will pay attention to common sense when you do start talking. Sheriffing could be rewarding when you got somebody a job or helped a wayward kid; but frankly, I don't miss it. These days, if I want to take my grandson fishing up in the Sweetgrass, I take him."

Rob Powell looked off at the glacier-capped peaks just west of his hayfield. "This is a beautiful place," he said, "and I intend to enjoy it as much as I possibly can in whatever time I've got left. I only wish that everyone else in the tribe could find a way to do the same."

U.S. Border Patrol Agent Cliff Heuscher, who grew up on a ranch just four miles east of Rob Powell's, had been recommended to me at the USBP headquarters back in Havre as a maverick agent with a reputation for speaking his mind. He was just coming off duty when I arrived at his trailer within sight of the official border crossing at Port Piegan.

I asked how he happened to join the border patrol. Cliff gave me a shrewd, appraising look from under his white cowboy hat. Then he grinned and told me that he'd attended high school nearby, in St. Mary, and had gone on to pick up a teaching degree at Western Montana State. He taught high school social studies for two years but didn't care for the bureaucracy in the public education system, so he defected from the teaching ranks to drive a cement truck. He worked as a range rider and then on a dude ranch. He was driving a wrecker when the local border patrol chief suggested that he apply to the agency.

"I didn't have anything better to do at the time so I took him up on it," Cliff said. "My first assignment was Yuma, on the Mexican border. That was an interesting place to work, with people illegally crossing into the States every night, and some of them desperate enough to do about anything to get in. But the fact is that being a border patrol agent is never anything but interesting, including up here on the Canadian border. These days another agent and I divide the area from Glacier Park all the way east to Del Bonita in the Sweetgrass. We each drive an average of two hundred and fifty miles a day, a lot of it through incredibly isolated country where you can and do run into anything from whiskey and cigarette runners to alien and drug smugglers."

I asked Cliff Heuscher what his most interesting case was. He crinkled his eyes and gave me another look from under the brim of his cowboy hat.

"Well, it wasn't really a case so much as it was an experience. When I was first assigned back up here, I took several days just to scout out the territory I'd be patrolling. I did it on horseback, and my wife came with me. We rode along the entire border from Glacier Park to the Sweetgrass. That was one of the very best trips of my life. We took topographical maps, which I used to make my own detailed maps of the border, where all the ranch gates and woods and streams and coulees and grasslands were. Any landmark at all that I thought might be useful to me. We met and talked to all the people who lived or worked along the border — ranchers, loggers, Indian families and white families, RCMP officers, everybody. It was the smartest thing I ever could have done, because the two keys to being a good agent are knowing your country and knowing your people."

I mentioned to Cliff that many of the Native Americans I'd met on my trip had said, "What border?" when I told them I was following the U.S.-Canadian line from coast to coast, stating that the border never existed before the arrival of the European settlers and, in their view, it still didn't. Cliff gave me a deeply skeptical grin.

I asked, "Do you think a local casino would cut down on smuggling?"

"A casino would provide local Native Americans with an income and employment," he said. "But what they have to realize is that in the gaming business, they're very little fish. The big fish are out in Las Vegas and Tahoe, many of them are in the Mob, and believe me, they move in fast. What's more, I think that there are far better employment opportunities for Native Americans than dealing blackjack at a casino, but — and here's a fairly radical idea — that Indians may very well have to move off the reservation to find them. One problem with reservation life is that while the natives certainly want to be independent, they also want government housing, health care, and educational benefits. Furthermore, plenty of people who are one-eighth or one-sixteenth Indian want the same benefits as full Indians. That can get complicated. Where do you draw the line? In my opinion, most natives would be better off if the reservations were done away with. Indians want to feel rooted. That's admirable. But I believe that they'd be much prouder of their roots if they could be independent and successful off the reservation. I'll tell you something else, too. You'd be surprised how many Native Americans privately agree with me."

There was one more lawman I wanted to talk to before crossing the Continental Divide. That was a Mountie — any Mountie. Who were these fabled red-coated policemen who had brought peace not just to the border country of the Canadian West but to far-flung territories all the way north to Robert Service's Yukon? Toward this admittedly romantic end, I decided to leave Montana temporarily and visit the RCMP office in Cardston, Alberta, on the old Whoop-up Trail.

The Whoop-up Trail was the haunt of drifters, scofflaws, hired guns, whiskeymen, wolfers, and all-around rascals — one of the most disreputable rogues' galleries ever assembled along the Canadian-American border. So I was amused to note, when entering Canada at Port Piegan the standard self-congratulatory plaque to harmony along "the friendliest border in the world." "This unfortified boundary line," the sign read, "should quicken the remembrances of the more than century-old friendship between these countries. A lesson in peace to all nations."

In 1896, Sergeant W. B. Wilde of the Northwest Mounted

Police was more interested in bringing law and order to the border country than in providing an international lesson in peace. Unfortunately, while patrolling the Pincher Creek area of what is now Waterton-Glacier International Peace Park, Wilde was bushwhacked out of his saddle and killed by one Charcoal, a renegade Blood Indian already wanted for murder. Charcoal was soon afterward caught and brought to justice by Wilde's fellow Mounties, who subsequently became renowned for their helpfulness to Indians and settlers alike.

Sergeant Ward Hoffman, out of the Cardston RCMP office on the Blood reserve north of Glacier Park, seemed to fit this enlightened tradition perfectly. Only in his watchful, level eyes, which had an expression of accumulated experience and the wisdom that usually comes with it, did he look older than any other twenty-eight-year-old. Otherwise he could have passed for, say, a Canadian Football League linebacker just out of college. My visit with him, though brief, was revealing.

Sergeant Hoffman began our conversation by telling me that the physical and academic training he'd undergone at the Mounties' training program in Regina was comparable to the FBI's. "In fact, the RCMP has a similar role in Canada to the FBI's in the United States," he said. "I personally like the challenges of the work, but there are plenty of them. Here in Cardston, for instance, the Blood Indian Reserve is part of a no-alcohol buffer zone that extends for seventeen miles north of the border, but there's a steady flow of illegal liquor coming in from the States. We're in constant touch with your U.S. Border Patrol people, and even so, it's a hard job to contain the smuggling.

"Our main accountability, though, is to the people we serve," he told me, and the words were scarcely out of his mouth when a receptionist put her head inside the door. "Crystal Snow Mountains is here to see you, Ward," she said. "Again."

"I'll have to ask you to excuse me," Ward Hoffman said. We went out into the waiting room, where I had an illuminating glimpse into exactly what he meant by accountability.

"Crystal, it's good to see you," he said in a genuinely warm voice to a young Native American woman standing by the receptionist's desk. Two toddlers, about three and five, clung to her legs, and there were tears in her eyes, one of which was badly bruised.

"I'm sorry to bother you, Sergeant Hoffman," Crystal Snow Mountains said, choking back a sob. "But he's done it again. This time I've had it. I'm not going back no matter what."

"I understand," Ward said calmly. "You don't have to go back."

He knelt down and smiled at the kids. "Hi, guys."

"Ma needs help," the older one said.

"I understand," Ward said again.

"What *can* he do?" I asked the receptionist as Ward led Crystal Snow Mountains and the kids into his office.

"Same thing he always does in a situation like this," she said. "Get her and the children into a shelter. Go deal with the husband if she wants to press charges."

She hesitated. Then she said, "Obviously, these situations don't have perfect solutions. But one thing's for certain. Sergeant Hoffman will do everything humanly possible to help that family. That's what the RCMP are here for."

Two Disappearances

Howard, why shouldn't I have a girlfriend like any other guy?

— My friend and fishing partner Nelson, earlier that summer

MIDAFTERNOON. Crowsnest Pass, Alberta. Up and up I go, the car straining on the steeper grades, shifting itself down into second gear and pulling hard. Except on the prairie, you're never far from logging country when you travel the border, and here in the Canadian Rockies the log trucks are carrying eighty-foot-long ponderosa and lodgepole pine logs, which hang out fifteen feet behind their trailers, travois-style. The logs sway back and forth precariously on every curve. DO NOT PASS. LOGS MAY SWING INTO YOUR LANE, proclaim big signs on the backs of these behemoths. I drop back a little farther.

Decades before the arrival of logging trucks in this country, gigantic coal deposits were discovered in Alberta's high-peaks border country. Miners sunk shafts deep into the interior of the mountains above Crowsnest Pass, overlooking the tiny town of Frank. Some of the garage-sized ovens where coke was once made to fire eastern steel mills still hunker down into the aspens and willows just off Highway 3 through the pass. The disused coke ovens have the same forlorn air of abandonment as the fire towers and rural drive-in movie screens and gutted schoolhouses and overgrown farms and collapsing railway stations and empty ore docks I've seen earlier on my trip. As for the once-booming town of Frank, it seems to have vanished entirely — and, in fact, it has. Eighty-six years ago, in one of the most horrific natural catastrophes in the history of the North Country, Frank, Alberta, disappeared from the face of the earth in a matter of seconds.

It was early one morning in 1907. Frank's miners had trooped up the mountainside from the village to work as usual. And as usual, they descended into the mineshaft high above the town and began their customary routine, digging ever deeper for the valuable coal deposits. Below in the town, it was also business as usual. The coke furnaces were fired up, the school and post office and company store were buzzing, the coal train came puffing up the spur line to be loaded. Suddenly a thunderous rumbling shook the entire region. An avalanche! Every man inside that mountain had heard disaster stories about miners being buried alive in their own tunnels, but luckily, most of the terrified men were able to dig their way out. When they reached the fresh air and sunshine, they rejoiced — until they saw, far down the mountainside, on the spot where their homes had stood, a great mound of stones. The entire village had been obliterated by a single chunk of limestone that had broken off the mountaintop and crashed down onto the town, shattering when it struck like an Old Testament judgment. The dislodged section of the mountain was estimated to have been five hundred feet high by sixteen hundred feet wide and more than a third of a mile long, and it killed sixty-eight people instantly.

The miners are gone today. The coke ovens slump smokeless into the mountainside. The scene of the avalanche is presided over by a single building: Kingdom Hall, the church of the Jehovah's Witnesses.

All my life I have been haunted by disappearances — of towns, farms, big woods, and people like Jake Blodgett, whose like won't be seen again. Somehow, the cataclysmic disappearance of the town of Frank made me think of my friend Nelson, at home in the Northeast Kingdom.

Nelson was sixteen when I met him, a big, wild, laughing kid who simply refused to go to school. "We don't know what to do with him," his father said to me the morning he first brought Nelson to see me. "We thought maybe, working with kids and all, you could help?"

As a social worker, I wasn't of much help to Nelson. He never held any of the dozen or so jobs I got him for more than a few weeks, or stayed more than a few months in any one place, or had another close friend that I knew of. He weighed close to three hundred pounds, had epilepsy, and at times would fly off the handle and take a swing at someone just for the merry hell of it. For years he'd been in and out of community group homes and foster homes. When he was out of school, which by then was all the time, he walked the roads picking up empty soda and beer bottles and panhandled on the streets of northern Vermont villages, where he was a perennial thorn in the side of the local police.

At the same time Nelson was something of a savant, with the vocabulary of a college literature professor, the wit of a standup comic, and a keen sense of irony extending even to his own handicaps. Like me, he loved to fish for trout, and after we became friends, we fished together for years. Then, just a month before this trip, I got a disquieting phone call from his father saying that the day before Nelson had gone fishing on the little stream beside his house and hadn't been seen since.

As I drove fast up to his place that morning, I thought that this wasn't the first time Nelson had disappeared. Once he'd lit out for Chicago, where he'd bunked in for a few months at a religious community. On another occasion he'd thumbed all the way to Texas. But in the past he'd called home every day or so and had touched base with me to report his progress and talk about fishing. This time he didn't have his antiseizure medication, and without that, he couldn't survive for more than a few days.

For the rest of the day, and the next and the next and the next, Nelson's family and I combed the woods along several of the border streams that he and I had fished together over the years. Phillis helped, and so did our son, Jake, home from college for the summer. We tramped through thick brush and bogs and forests, shouting Nelson's name at the top of our lungs. Nothing. We tried to mobilize the state police and neighbors to form a search party, but no one else was concerned enough to help. A week went by. Two. And still no trace of my friend and fishing partner.

The police tried to reassure us. Hadn't Nelson done this a hundred times before? No, his father and I said on our daily visits to the barracks, not like this, without contacting anyone, without taking along his medication. Next the police dredged up vague reports that he'd been sighted twice in the past few days. No, we told them. We'd checked both "sightings" carefully; neither held a drop of water. Eventually the cops dispatched a couple of border patrol officers with dogs to one or two of the brooks Nelson conceivably could have fished. Nothing.

"He's out there," I kept telling Phillis. "In the woods somewhere."

I couldn't stop thinking of Nelson — alone, helpless, without medication, maybe dead. His father and I ranged farther, walking streams miles from his house. It was as though Nelson had been swooped up by his ever-present extraterrestrial companions, whom he'd described to me in great detail that first morning I'd met him and many, many times since.

As I crossed Crowsnest Pass in Alberta, I thought with a vast hollowness of how we all had searched for three weeks, futilely begged the police and neighbors to help look, and hoped against hope that the phone would ring and Nelson's laughter would come hooting over the line from California or Florida or wherever.

But when the phone did ring, the sober voice of a local police sergeant informed me that Nelson had been found dead in a brushy section of the tiny brook near his house, a place we'd searched within two hundred yards of ten times but had somehow overlooked. Apparently he'd had a seizure and had fallen into the water — leaving me with the saddest North Country story I know.

8:00 P.M. On the Montana–British Columbia border at full dark. It's cold with snow on the way, if not tonight then very soon. Summer in the Rockies has disappeared. Time to go fishing.

A Fishing Idyll

Then the universe stepped on its third rail. The wand jumped convulsively as it made contact with the magic current of the world. The wand tried to jump out of the man's right hand. His left hand seemed to be frantically waving goodbye to a fish, but actually was trying to throw enough line into the rod to reduce the voltage and ease the shock of what had struck.

— Norman Maclean, *A River Runs Through It*

I rented a motel room in Eureka, Montana, to use as a base camp, and dawn found me on a likely-looking bend on British Columbia's Elk River, stringing up my fly rod. For the next several hours I caught and released rainbow and brown trout from twelve to fifteen or sixteen inches long, one right after another, on a yellow grasshopper fly. The trout were feeding as if they knew that the grasshoppers would soon all be gone, and there was no trick at all to catching them. Find a riffle or a deep run at the head of a pool, cast across the current, let the fly float downstream, and as often as not I'd have a fish on before I knew it.

Around noon I drove back across the border to Montana's Flathead River, which Phillis and I had fished on our trip west in 1969. There I repeated the morning's performance, fishing happily on into the afternoon, even pushing up a brawling little tributary and taking a few brook trout and cutthroats. Although I hadn't yet hooked a really big fish in the category of that lunker that takes Paul Maclean down the rapids in *A River Runs Through It*, I was having a time for myself catching and releasing gorgeous, acrobatic trout of four different varieties — how Nelson would have loved doing this, I thought; how my father and uncle would have loved it.

In the late afternoon I found myself sitting on a rock in the middle of the Flathead, letting my fly drift downstream and daydreaming about my family and friends and the rivers that had run through my own life. It was a fine thing to have an interlude for reflection on these western trout rivers at the end of the Outlaw Trail; and as evening approached it occurred to me, inveterate fisherman that I am, that it would also be fine to catch one really big trout.

It was much cooler now that the sun was off the water, and with the natural grasshoppers all in bed for the night, the trout were no longer coming for my imitation. A Muddler minnow streamer might attract a big fish, but I had an even better idea. Earlier that afternoon, in a deep pool where the icy spring-fed tributary in which I'd caught the brook trout emptied into the Flathead, a very large brown trout with a bright yellow belly had flashed once at my fly, then retreated under the bank beneath a cottonwood and refused to come out again. Now I thought I knew what he wanted.

On my way back up the river toward that pool, I turned over several fallen logs along the bank until I found what I was looking for: a big, lively angleworm. I couldn't help thinking of Norman Maclean's good-for-nothing boozing brother-in-law bringing his whore and his Hills Bros. coffee can of lowly garden worms to a blue-ribbon Montana fly-fishing stream; but remembering the Allagash fishing guide Harry Hughey, who sometimes "carried his flies in a gallon pail," and my father, the finest fly fisherman I've ever known, who taught me how to fish by rigging his fly rod up with a worm and sinker, I was far more amused than chagrined by what I was about to do.

By the time I arrived at the pool, the sky was the same deep purple as the sage on the far hills, and it was cool enough for my lumber jacket, back in the car half a mile away. But there was still plenty of light left to thread the worm once through a number-ten Eagle Claw hook I kept in my fishing vest for just such an exigency. Careful not to spook the fish this time, I flipped the worm into the river where the creek spilled in, just as my father had flipped a worm into the culvert pool where I'd caught my first trout, on the same fly rod I was using now. I let out the line as the bait rode down the current, sank, and worked its way deep in under the cottonwood on the bank. There the worm sat, and so, on a nearby log on the sandy shore, did I, with the beautiful split-bamboo Orvis Battenkill fly rod my father had given me propped up in the notch of a forked stick stuck into the sand.

I was sure that the big fish was still out there. The mouth of the creek was a perfect feeding spot, and although my fly or my profile had put him down earlier that day, this time I'd approached the pool with great stealth. I felt certain he'd be unable to resist that fat, wriggling worm, and sure enough, a minute later the tip of my pole began to bounce like a divining rod over an artesian well. The line, where it entered the water, was cutting upstream, toward the mouth of the creek. Glory be, I thought. Here was my fish. I set the hook.

It didn't move my way at all, and unlike Paul Maclean's mammoth rainbow trout in *A River Runs Through It*, it disdained aerial displays for a deep strong battle on the gravel bottom of the pool, jumping just once and just high enough for me to see that this was no tail-walking crimson-sided rainbow but a big hook-jawed German brown, no doubt the same trout that had swirled at my fly earlier that afternoon. And once again I felt connected through a fighting trout to a flowing river and the twilit northern countryside it flowed through — though I was already hoping that this fish was hooked lightly enough so I could release it.

The trout and I were both lucky that evening. When I eased it up by my feet, I saw that it was just barely hooked in the tough cartilage in the side of its jaw. It was no leviathan but certainly a good fish, a big male upward of two feet long and weighing perhaps five pounds, with a lemon-yellow belly and vermilion spots the size of dimes on his sides, and when I held the bent rod high over my head, running my other hand down the leader and twisting the hook free, the fish turned and swam unhurriedly back into the depths of the pool, unharmed.

Part Five

FRESH STARTS

The Last Best Place

It was a slow day, a slow town, and we finally met a realtor, Ross, who was delighted with our plight, and he began telling us about a wild, magical valley up on the Canada line over near Idaho. Yaak, he said, wasn't really a town — there was no electricity, no water, no paved roads — but a handful of people lived there year round, sprinkled back in the woods and along the Yaak River.

— Rick Bass, *Winter*

4:00 A.M. Eureka, Montana. This is big timber country, and long before daylight I'm awakened by the growl and rumble of idling logging trucks. It's pitch dark outside, but parked in front of the twenty-four-hour café next to my motel are a dozen or so semis, outlined by their red and green running lights. Most are hauling empty doubles, with the shorter second trailer or "pup" folded neatly up onto the big trailer, wheels in the air, piggyback style. After my fishing jaunt the day before, I feel like a million dollars, and a few minutes later I'm ready for a cup of cowboy coffee and some trucking conversation. I don't have to wait long for either. Three drivers about my age, wearing checked shirts, belts with ornate buckles, jeans, boots, and cowboy hats, make room for me at a window booth. The waitress sets a pot of steaming hot coffee down in front of me, and for the next twenty minutes no one at that booth utters a single syllable that doesn't pertain to trucks or trucking.

4:30 A.M. The drivers are drinking their coffee from personalized plastic mugs upward of a foot high and embossed with their names — Ross, Harvey, and Jonesy. Ross, a heavyset man in his late forties with a baritone voice and an easy, engaging style of conversation, does most of the talking. He says that the truckers in the café this morning will be picking up lodgepole pine logs just over the border in British Columbia and hauling them down to the big mills in Whitefish and Libby. He says that most local log-truck drivers work sixteen hours a day, counting loading and unloading time. They try to make at least two round-trip runs, though depending on how far into the woods they have to go to pick up their logs, they occasionally squeeze in a third. "Two runs is fine for me most days," Ross says. "Take yesterday. I could've squeezed in a triple. Instead I came home early and took my boy and girl up back of the house and we built a tree stand for bow hunting. After we got it up, we sat there together and watched four deer come out into the clearing below. So I lost an extra run. So what? This country isn't going to last forever. Neither, for that matter, am I. I want to enjoy it while we're both still here."

4:45 A.M. "Is this territory up here what's known as the last best place in Montana?" I ask.

Ross: "Well, Eureka's fine country. So's the Sweetgrass. But the last best place in the entire lower forty-eight is over in the Yaak, to the west of here. Before you leave Montana, drive up there on the wilderness road. You'll see what I mean."

5:30 A.M. Up I go, ascending the twisting wilderness road by way of thickly wooded switchbacks, as the sky in the east turns fiery orange and the sun rises over the Rockies behind me. From the divide at the top of the mountain, the Yaak Valley spreads out as far as I can see. Much of the countryside has been logged within the past decade, and the new young trees all around me shine a deep emerald in the sunrise. Suddenly a big mule deer crosses the road. It's moving fast, its front feet landing close together the way mule deer run, appearing to hop like a kangaroo. When you see a deer cross the road anywhere in the North Country, there's usually another one nearby, so I slow to a crawl, and sure enough, a second comes bounding across the road close behind the first, it too bouncing as though on springs. Around the next hairpin I encounter two whitetail deer, which don't bother to run at all. In the next five miles I count twenty-five deer.

* * *

8:00 A.M. I know from Rick Bass's superb memoir *Winter* that there's some sort of hamlet nearby, a tiny village in the wilderness consisting of one store and two wonderfully named saloons: the Dirty Shame and the Hellroaring. After driving another ten miles, I find them. Across the road from the Dirty Shame Saloon and just up the way from the Hellroaring, I stop at the local mercantile, as general stores are called throughout the Northwest, and meet Jack Pride, a hunting guide and fifth-generation Montanan. I ask him if a new paved highway just being driven up from Libby is apt to open up the last best place in the lower forty-eight to an influx of newcomers, and if so what that will do to the hunting.

He laughs and says that when his grandfather first came into the Yaak sixty years ago, *he* was a newcomer. "He was working at the Libby Mine when they struck gold on the Yaak River. When the gold played out, he stayed on and homesteaded. Shot a deer now and then, grew his own potatoes and turnips — root vegetables are about all you can grow up in the Yaak — caught all the trout he wanted. My dad lived the same way, off the land. That's how I grew up, and when I was a kid this was still probably the last best place anywhere, for natives and newcomers alike. No, sir. It isn't the newcomers we mind. It's outsiders who try to tell the rest of us how to run our affairs and use our land and live our lives.

"I'll give you an example. A lot of people are opposed to hunting, and when it's done irresponsibly so, frankly, am I. But very few Yaak families ever take more meat than they absolutely need these days, and I make damn sure my clients never take more than the legal limit. For most of them, just getting out in the wilderness with the elk and deer and bear, maybe even seeing a mountain lion track, is the thrill of a lifetime. We never leave a wounded animal in the woods to die. I make sure I know what sort of shots my clients are before we go out. Point is, we know how to handle our wilderness responsibly, and intend to. That way, the Yaak will be the last best place for a long time to come — for everybody."

Boundary County

The destruction of the Weaver family, and Randy Weaver's subsequent exoneration on Federal conspiracy charges, is today the parable by which many living off the grid in Idaho and elsewhere understand the government's relationship to the American family.

— Philip Weiss, "Off the Grid," *New York Times Magazine*, Jan. 8, 1995

Idaho, with its utterly distinctive panhandle jutting up to the Canadian line between western Montana and eastern Washington, was usually the first state I reached for when I was a small boy fitting together my wooden "Map of the United States" puzzle. Like Texas and Florida, Idaho's unique configuration made it stand out.

The end of Idaho's "handle," Boundary County, contains some of the wildest frontier country left in the United States. During the past decade this wilderness has attracted thousands of contemporary individualists in search of elbowroom and freedom from what they regard as intrusive government regulations, including Christian fundamentalists, survivalists of every stripe, and white supremacists. Almost exactly a year before my visit, a desolate northern Idaho mountaintop was the site of the much-publicized shootout between the survivalist Randall Weaver, wanted by the FBI on conspiracy charges, and a posse of federal agents. During the course of the gun battle, Weaver's wife and fourteen-year-old son were killed, and a deputy U.S. marshal was also shot to death. Weaver was subsequently cleared by an Idaho jury of all charges but failure to appear for an illegal-weapons hearing, and his case became a nationwide rallying cause for individuals opposed to government regulations of all kinds, as well as a focal point for groups opposed to gun control.

I was more interested in the broader appeal of Boundary County as a frontier for newcomers wishing to make a fresh start.

Having witnessed the phenomenon of Northeast Kingdom communes that flourished for a time during the 1960s and '70s, then mostly went the way of northern Vermont's abandoned hill farms, I wondered how contemporary off-the-grid newcomers in the panhandle of Idaho were doing, and I wanted to meet some of them.

Bonners Ferry, the county seat, sits compactly inside the elbow of the Kootenai River where it bends north to head into Canada. Here in this valley in 1808, the British explorer and mapmaker David Thompson was saved from starvation by Indians who fed him dried fish and bread made from moss. After regaining his strength, Thompson was able to make a fresh start of his own and forge on to the Columbia River and the Pacific. I couldn't help thinking, as I drove into Bonners Ferry nearly two centuries later, that the once-glorious profession of explorer, in the tradition of adventuring visionaries like Thompson and Meriwether Lewis, had in my lifetime become anachronistic.

I settled for dipping into the county courthouse, where a clerk recommended that I see Bruce Whitaker, Boundary County's sheriff, whom she described as a highly respected young man who should have been allowed to handle the Weaver situation himself, without outside interference. If he had, she assured me, the whole tragedy could probably have been avoided.

Intrigued, I stopped by the sheriff's office and caught Sheriff Whitaker between appointments. He was reluctant to discuss the Weaver case, which was still under investigation, but he didn't seem overly concerned about most of the other controversial newcomers to Boundary County, the survivalists and back-to-the-land folks. "Basically, geography is our main problem here," he said. "Boundary County consists of eighteen hundred square miles, and it's impossible to patrol it all adequately. It shares sixty miles of border with Canada, most of it very rough terrain. It's the only county in the U.S. that borders two other states and a foreign country. Right now I have twenty-three full-time patrol staff and could use double that for domestic abuse cases alone. We do everything we can to bring domestic abuse to light, stop it, educate the public about it, prevent it. But it's still our number-one problem and almost certainly the number-one social and law enforcement problem throughout the North Country."

After this sobering conversation, I drifted down the street to the local bookstore, Bonner's Books. The young proprietor, John O'Connor, had grown up in Seattle and made a fresh start here in Bonners Ferry about a decade ago. How, I wondered, had he become a book man in an area with a reputation for attracting cultists who valued their home arsenals far more than their home libraries?

John smiled. "I got into the business myself for the same reason most booksellers seem to, which is that I like to read. Back in Seattle I worked at the university bookstore. But in a big city, unless you have huge amounts of capital, you don't have a snowball's chance of owning your own store. Finally I sort of dropped out of the system and came to Boundary County as a refugee, like hundreds of other young people. I got a job working in the hops fields up by the border. During the off-season, I did a little of everything. Worked on hog farms, grew a garden, lived off the land. But I needed a steady income, so eventually I got a job at the store here, working for the previous owner. Gradually I became integrated into the community. When the owner decided to sell the store, I got some help from the local bank, and one morning I woke up and found myself the owner of a bookstore."

"If you were going to write about making a fresh start in Bonners Ferry and Boundary County, what would you say?"

John thought for a minute. Then he said, "I'd find an old-timer to talk to. Someone in touch with the history of the community. The longer I live here, the more I think that in order to understand what's going on in Boundary County today, you have to know how the area was settled and what the frontier here was like one hundred and fifty years ago, when everybody in this territory was making a fresh start. To find out about that, talk to Walt Maki, up the street at the hardware store. He can tell you all about the history of Boundary County — *if* he likes you, that is."

Walt Maki's hardware store was a long, narrow cavern of a building that seemed to reach halfway back to Montana. Everywhere were bins of nails, drawers of screws, shelves piled high with all the tools and materials you would ever need to keep a house in good repair. Twenty feet into the store the daylight faded out altogether, and I entered a crepuscular netherworld dimly illuminated by a few widely spaced naked overhead bulbs. Way in the back I heard someone rummaging around. It turned out to be Mr. Maki, a rugged-looking man in his seventies, who seemed determined to ignore me even after I introduced myself.

"John O'Connor down the street at the bookstore said you could fill me in about the history of the area," I said.

"Depends what you want to know," he said gruffly.

"For starters, I wondered how Bonners Ferry got its name?"

Still rummaging, Walt Maki said, "Well, how do you suppose? Old Man Bonner had a ferry on the river here. Ran it for years."

"What was here before that?"

"Wild Horse Trail to Canada. Smuggling path."

"Were your ancestors smugglers?"

I'd hoped Walt Maki might laugh, but I should have known better. "My people were hard-working Finns," he said. "Like me."

At this juncture almost anyone but a nosy writer would have taken the hint and moved along; but apparently Walt Maki had been thinking about my interest in the town's history, because suddenly he straightened up and said, "You want to see the first bank of Bonners Ferry?"

"Sure."

He jerked his head backward, directing my gaze to the biggest floor safe I'd ever seen, bulking up out of the dusk behind him. I could just distinguish the words BONNER MERCANTILE, LTD. in faded letters on its door.

"That safe goes back to Old Man Bonner's days," Walt said. "He had a trading post here, and the trading post was the town and the safe was the town bank. Folks had any valuables, this is where they kept 'em. For a long time his safe was the only bank in Boundary County."

"How come it says Ltd. instead of Inc.?"

"Oh, Bonner and most of those other early boys were British. Same difference, though. All it meant was if the mercantile failed, Old Man Bonner's creditors could sue him and latch onto it, but they couldn't take away his home or personal property."

"Was Bonners Ferry a wide-open town?"

"You bet it was. And remained that way right up through the 1920s and '30s. I can remember when there was a whole string of whorehouses over along the river. We called that part of town 'the cribs,' because the whores lived in tiny little one-room holes like cribs."

"Your sheriff says Boundary County's still a pretty wild place."

"Yes, but times have changed, and in some ways for the worse. Here's an example. Back during Prohibition, there was a black fella over by the local reservation named Ed Black, oddly enough, that made moonshine. He was the only black man in the county and he made number-one moonshine and he was a number-one violinist. Ed minded his own business and provided two valuable commodities to the community, booze and music. So nobody bothered him until the federal men, the revenuers, hit town. They nosed around to see who was making shine and came across old Ed. Brought him down to the courthouse and hauled him up in front of a grand jury. When the prosecutor from Coeur d'Alene came up, he said to the local sheriff, 'What do you want me to do? Is this fella Ed Black a menace? Should I put him away?' And the sheriff, old Sheriff Broder, said, 'No, he makes the best moonshine for a hundred miles in any direction and he's the best musician in Boundary County and he minds his own business.' So the prosecutor threw the case right out the window.

"Now," Walt Maki said, looking at me hard. "That's exactly what should have happened with Randall Weaver but didn't. The feds waltzed in here thinking they were God and they shot up Weaver's house and killed his wife and boy and that wasn't right. You can't murder a man's family because the fella missed a court appointment. The law's supposed to protect everybody. Old Sheriff Broder who spoke up for Ed Black knew that. Today the feds have way too much power and abuse it. It happened here, it happened in Waco, and it's wrong. You think about it."

When I stopped for gas at a ramshackle mercantile just south of the Idaho–British Columbia border, I spotted a shirtless, skinny guy in his thirties sporting swastika tattoos on his arms and packing a revolver in a shoulder holster. He was just getting into a pickup adorned with bumper and windshield stickers proclaiming AMERICANS — YES. FOREIGNERS — NO.

On my way into the store I paused by the pickup's open window. "Hot day," I said.

The shirtless driver stared at me.

"I wonder if you could give me some directions," I said. "I'm a writer from Vermont and one of the things I wanted to do out here was talk to a survivalist."

"Well, you're talking to one. Me and my wife and kids have lived up here off the grid for two years."

"Where do you live?"

"Out there," he said, pointing at the mountains up by the border.

"Do you and your family live off the land?"

"More or less. But the point isn't what we live off from. It's what we live *away* from."

"What do you live away from?"

The guy leaned toward me slightly with his head partway out the window, his shoulder holster swinging a few inches out from his bare chest. "Jews," he said. "Queers. Niggers. Nonbelievers. Schools. Government. Taxes. Electricity. The devil. Spics. Writers like you."

Then he withdrew into the cab like a rattlesnake pulling back into its den and tore out of the parking lot with his tires spinning, leaving me speechless for the first time since leaving home.

"Who was the guy with the pistol and swastikas?" I asked the woman behind the counter at the mercantile.

She looked around to make sure we were the only ones in the

store. Then with a slight Mexican accent she said, "Asshole from up in the mountains by the border."

"What do they do up there?"

"Take target practice, I guess. We hear a lot of shooting. They call themselves survivalists, but really they are just a bunch of wacko Nazis."

"Are they dangerous or just crazy?"

"Both," she said. "In my opinion they're both. My husband and I, we first start coming up here from Mexico ten years ago to work in the hops fields. Well, we like it. Is beautiful country, yes? Five years ago we become naturalized and rent a small place. My husband, he still works in the hops and does some logging. I have two jobs, this one here and another at a store in Bonners Ferry. We save for our own log truck. We even put away something for the kids' college. Now these locos are coming in here with their swastikas and guns. I say to my husband, we move all the way up here almost to Canada, we still can't get away from it. Get away from what, he says. He don't take the locos as serious as me. From racism, I say. From hate. Is everywhere."

"These guys with their guns and hate slogans. You think they may not know what they're saying?"

The woman shook her head. "They know what they are saying. You had just better believe they know what they are saying. Look. Idaho is a good place to live. Good for kids, for a new start. But those Nazi hombres? They are bad people with bad ideas. Believe me. They know *exactly* what they saying."

"Are you going to stick it out?"

She looked at me thoughtfully, maybe wondering who I was. But before I could tell her, she nodded. "We stick it out," she said. "We fight them, how you say? Tooth and nail, if we have to. We did not come here just to run away again when things get tough. Yes. We stick it out."

Notes from the British Columbia Border

It was an exuberant, staccato summer. Luck and events and kindnesses melded with one another. I rushed along eagerly, without any special introspection, just putting down what I found out.

— Edward Hoagland, *Notes from the Century Before: A Journal from British Columbia*

2:00 P.M. As I approach the Porthill, Idaho, customs station, I find myself driving into a dense haze rising from some flats beside the river, where the recently harvested Budweiser hops vines are being burned off for the winter. The sun is a pinkish halo through the thickening smoke, which has a heady, sweet scent, like exotic pipe tobacco — like burning maple leaves at home in Vermont.

2:15 P.M. At the customs station Agent Walt Worley tells me that his grandfather, one Bear Worley, came to the high-peaks country of northern Idaho to make a fresh start around the turn of the century. "His intention was to mine silver, and he wintered over the first year in his own mine tunnel. But he was always more interested in hunting than in mining. After a while that's all he did. He lived to hunt, especially bears. That's how he got his name. Porthill's also famous as the site of the first major drug bust along the Canadian border west of the Continental Divide, back in 'ninety-two."

Bear Worley's grandson grins at my puzzlement. Surely there had been major drug arrests along this stretch of the border before 1992? This must have been something colossal. A truckload of heroin, maybe?

"Nope," Agent Worley says. "Fifty pounds of opium, headed south into the States — in *eighteen* ninety-two."

2:45 P.M. In Creston, British Columbia, just over the border from Porthill, I stop for gas. "Did you bring all this smoke to

Creston?" the blond woman at the register demands. "From those damn hops fields?"

"No, ma'am. It looks like it's been blowing up here all day."

"It has. You know something? It seems to us like the old Budweiser Company only burns off their hops fields when the wind's blowing north."

I laugh. "Well, I didn't bring the pollution. I'm a writer from Vermont."

"No, you aren't."

"I'm not a writer?"

"You may be a writer. You're not from Vermont. No way."

"Why not?"

"You don't talk like a Vermonter."

"How do Vermonters talk?"

"You know. Like Larry, Darryl, and Darryl, on the old Newhart show."

Now we're both laughing; but when I tell her that Vermonters don't talk like Larry, Darryl, and Darryl, and in fact I've never heard anyone talk exactly like them, she's skeptical, or pretends to be.

"Where're you headed?" she asks.

"Up into the Nelson Mountains."

"Watch out for avalanches."

"Sure."

"I'm serious. If you're going up there, watch out. They have several avalanches a day in those mountains."

Right, I say to myself. Avalanches. Like the Frank Slide. Scare the rustic Vermonter.

3:00 P.M. Here in the West I sometimes find it easier to stay close to the American North Country by dipping up into Canada and hugging the north side of the border, as I did in the Milk River badlands of Alberta and again at Crowsnest Pass. It's just midafternoon, and Route 3 over British Columbia's Nelson Range seems right for me, avalanches or no. Near the top of the pass, I come to a blinking caution light with a gate just beyond. The gate is open, but a large sign warns me to slow down for avalanche control work ahead. Great, I think, but I continue until, just as I crest the divide and start down the precipitous switchbacks on the western side, I'm waved to a halt by a good-looking young woman in a green hard hat, a fluorescent orange jacket, work boots, and blue shorts. Far below, a bright yellow bucket loader is inching its way up the highway, which is steeper than any road I've traveled since leaving home and totally devoid of guard rails. Just beyond the bucket loader Route 3 bends sharply north to hug the mountainside, and the gravel shoulder ends in space, over a drop-off plunging to infinity. I feel like a novice skier at the top of an Olympic-sized jump.

"Road construction?" I call out to the young flag woman.

"No," she says merrily. "Avalanche control. See those guys up there?"

She points high up at the escarpment above the bucket loader, where some tiny figures in green hard hats and orange vests are watching from the brink of a ledge. "They just dislodged a boulder that was going to come loose anyway after another good rain," the flagwoman says. "See it down there by the loader? It came to rest on the shoulder."

I do see the boulder, which is as big as a Volkswagen, perched right on the brink of the precipice below me. With amazingly delicate precision for such a large piece of equipment, the yellow loader slides the foot-long steel teeth of its bucket in under the rock, picks it up to about cab level, rumbles forward a few feet, and dumps it over the edge of the cliff. Near the bottom of the chasm, hundreds of feet below, the stone takes a great erratic bounce off a ledge, plows through a stand of firs like a tornado, snaps a medium-sized tree right in two, and disappears from sight.

Surviving in Survivalist Country

I don't know how I missed that grouse this afternoon, but you know something? I'm actually glad I did.

— Me, to Phillis, a month after returning from my trip

I should preface my account of what happened the next day by acknowledging that I was undoubtedly still jumpy after my encounter with the pistol-toting supremacist. Maybe my jitteriness explains a great deal of the following — I don't know and never will. It was a bad experience, however, far and away the worst of my trip.

Having visited with a Maine game warden, a Minnesota taxidermist, and several North Country guides from New England to Montana, I very much wanted to go into these big western woods and meet some big-game hunters on their own terms — see their territory and spend at least part of a day with them. Like other North Country natives of my generation, I'd hunted most of my life (though not nearly as avidly as I fished), and although I had long been an advocate of strict gun-control laws, I had no objection to hunting when it was done responsibly — the way Jack Pride over in the Yaak did it, for instance. Like fishing, hunting has been a way of life in the North Country for as long as people have lived here.

The morning after I crossed the Nelson Range, I got up with the birds, packed a lunch, and hiked into the high-peaks region of northern Idaho from the British Columbia side. In the middle of the morning I crossed the Vista, the twenty-foot-wide cleared strip marking the border, and from there I followed a brook high up into a pass, planning to hook up with some bow hunters I'd met in a diner back in Bonners Ferry. The hunters had told me to look for their camp at a fork in the brook about six miles beyond the border, and since I'd brought my fly rod, I fished my way up the stream along the trail, catching plenty of pan-sized cutthroat trout on my reliable grasshopper fly.

It was another brilliant early-fall western day, a great day for a hike or a fish or a hunt; but somehow, though the directions the hunters had given me were clear enough, I never did locate their camp. The brook forked several times, and maybe I went up the wrong branch or missed their tents set back in the trees. Regardless, I wasn't equipped to stay out in the woods on my own, so in the middle of the afternoon, having covered the six miles and then some, I started back downstream. I estimated that if I walked straight back to the border and down the trail without stopping to fish again, I'd be out by dark with an hour to spare.

I soon discovered that I'd fished considerably farther upstream than I'd realized. By the time I reached the Vista, which was still at least three miles from the trailhead, it was five o'clock. At this rate, I'd be lucky to make the car by dark.

The hunter was standing just across the border on the British Columbia side, looking at me through a small pair of field glasses. When I'd first emerged into the clearing from the Idaho side, where I'd stopped briefly to reconnoiter, he hadn't been there. Nor had I seen him step out of the woods. He was dressed in a bow hunter's green-and-black camouflage outfit: camo shirt, pants, hat, and boots. His bow was wrapped with camouflage tape and his face was painted with green and black stripes. Even his hunting gloves were green and black. Only the feathers of the razor-sharp broadhead hunting arrows in his quiver were uncamouflaged, and those were a bright blood-red. Still, it wasn't his appearance but rather his materializing so silently and suddenly that startled me.

"Hello!" I called.

He nodded. Though it was hard to tell with all that paint on his face, he looked to be in his early forties. He was about six feet tall and in good shape, and though nothing about him physically really stood out, I was positive that he wasn't one of the four hunters I'd met in the Bonners Ferry café. When I told him that I was on my way down the trail to my car, he said he was due back at the campground where he was staying with his "mom and dad." That struck me as an engaging, if somewhat odd, way for a man of his age to refer to his parents in a conversation with a stranger, but I didn't think much about it at the time. Immediately afterward he said he'd walk out of the woods with me, which in fact is a very odd thing for a hunter or fisherman to say to a stranger — a slight but definite breach of the wilderness protocol that I learned at an early age from my father and uncle — who taught me never to infringe on anyone else's camp or fishing beat or deer stand or personal space until I'd been invited to.

We started down the trail together on the British Columbia side, still following the brook. Here in the deep woods, away from the cleared strip marking the border, it was already getting dusky, though there was still plenty of light left in the sky. I made a little desultory conversation, talking about the fishing and the run of fine weather, but I got little more than murmured replies from the archer. And, though we walked side by side on the trail, he didn't look at me when I spoke.

After twenty minutes of this, he suddenly pointed at a rippling ring on the surface of a nearby pool in the brook and said, "There's one!"

Did the bow hunter want to watch me catch a fish? Well, fine, I'd be happy to oblige him. I sneaked up on the pool so the trout wouldn't see me, tossed my fly out where we'd seen the rise, hooked the fish, and brought it flopping onto the bank. It was just a little cutthroat, maybe eight inches long, but fat and sassy and very pretty, with gleaming silvery sides and bright scarlet gill slashes just the color of the bow hunter's arrow feathers. I took it off the hook and turned around to show it to the hunter before releasing it, and I will be damned if he didn't appear to be scrutinizing me through his camo field glasses again — from less than ten feet away!

Jesus! I almost said it out loud. This business with the field glasses gave me a chill. But the hunter quickly put them away and affected an interest in the trout, or maybe he really was interested. Maybe he'd been using the binoculars to watch the fish take my fly off the water or to glass the steep, wooded ridge across the brook for an elk. And maybe it was the green and black paint on his face in those dim woods that made his eyes look spectral and macabre to me; and maybe my growing discomfort with the weirdness of this whole encounter was entirely the result of my conversation with the armed survivalist outside the Boundary County mercantile the day before.

Maybe.

I made a quick decision. "I guess I'll fish back down the stream to my car from here," I said. "See you later."

"I'll go back on the trail," he said after a brief hesitation. "I've got to get back to my folks."

And he headed off without further comment.

Everything about the situation still felt wrong to me, though I couldn't put my finger on a reason. I don't consider myself unusually suspicious by nature, and if I had to run into a nut, I'd certainly rather meet him in the most isolated region of the Canadian border than in a dark alley in a big city; in the woods, at least, I'm at home too. It did not occur to me until later that the hunter might have been afraid of me; that could have accounted for his reticence, but I don't really think that was the case.

So I waited a couple of minutes, and then I did something I'd never done in all my life. I deliberately followed another man through the woods to see what the devil he was up to. I slipped down the trail around one bend and around another — and then, shades of *Deliverance*, James Dickey's horrific tale of mayhem in the southern wilderness, I saw him, a couple of hundred yards away, standing off the trail behind a fir tree, watching the brook and waiting. For a deer or an elk? For me? I had no idea, but I realized that I was no longer merely unsettled. I was scared.

I watched him for a minute or two, hoping against hope that he wouldn't turn his head and look up the trail and see me. Okay, I finally said to myself. Enough of all this. *Do* something.

Very stealthily, I eased back out of sight. My heart was beating fast, but I knew exactly what I was going to do. I hotfooted it back up the trail in the direction I'd been coming from for five minutes or so. Then I crossed the brook, climbed fast up the steep, thickly forested slope, and started rapidly back down along the top of the ridge on the opposite side from the trail, paralleling it and the brook. From time to time I caught a glimpse of water, several hundred yards below me, gleaming through the quickly darkening woods. I stopped once or twice to catch my breath and study the terrain, but saw no sign of the hunter, who presumably was still on stand where I'd last seen him. When I was sure I'd passed the trailhead, which was opposite me on the far ridge, I crossed the brook again and walked back up the woods road in the twilight to my car. Next to it was a new white pickup with Idaho plates and an empty gun-and-bow rack above the back window of the cab. I didn't linger to find out if it belonged to my hunter. Twenty minutes later I was back on a paved British Columbia highway, as relieved as I could ever remember being.

The Upper Columbia and Bill Bingham

Like many Scottish ministers before him, he had to derive what comfort he could from the faith that his son had died fighting.

— Norman Maclean, *A River Runs Through It*

The Columbia River is to the Pacific Northwest what the St. Lawrence is to the Great Lakes drainage basin. It is America's largest river flowing into the Pacific and, with a length of more than twelve hundred miles, the fourth longest river on the continent. I crossed it with a surge of excitement the next morning at Northport, Washington, just south of the British Columbia border, where I stopped at a pulloff and clambered down the steep bank to talk to a fisherman. He was a man about my age, in a red lumber jacket almost identical to mine, and he was sitting on a log with his rod propped on a forked stick, smoking a cigarette and working on a quart bottle of Mountain Dew. Were there trout up in the border reaches of the Columbia? I asked.

The fisherman took a pull at his Mountain Dew, lit another cigarette, snorted, and informed me that there certainly were not, or salmon either, though at one time both species had thrived in the upper Columbia. Since the big dams had gone in downriver, however, this stretch of the river was full of "trash fish like bass." This made me laugh out loud, and since I wasn't fishing myself, I asked if I could stick around for a while. He nodded bleakly at the log he was sitting on.

The fisherman's name was Bill Bingham. Originally a native of the area, Bill was a newspaperman who'd written for and edited small dailies and weeklies in a long string of North Country towns from Montana to the Pacific. He had a cabin nearby and had been coming here to fish for more than thirty years. Bill was a big, broad-shouldered, chain-smoking man, with short white hair and a pugnacious stare, as though everything his blue eyes lighted on offended him in one way or another. Naturally, I liked him immediately.

When I mentioned that I was interested in Westerners making fresh starts, Bill said he'd recently made one himself, in honor of his fifty-first birthday. Sixteen days and — checking his wristwatch — twelve hours ago, to be exact, he'd taken a leave of absence from his newspaper and come up to his cabin near the Canadian border to dry out. "I was drinking a bottle of Seagram's a day before I left, and it was flat-out killing me. So here I am, bait fishing with a bobber on a trash-fish impoundment, just to keep from drinking."

"What's wrong with that? This is still a beautiful place to fish, a beautiful river."

Bill snorted. "Do you call this a river? I don't. This is a goddamn glorified millpond. Look. The Columbia used to be one of the finest salmon rivers on the face of the earth. My great-grandfather and my grandfather both made their living catching salmon in drift nets, put 'em in the river up here and they'd float down to Kettle Falls, where they'd take 'em out so full of fish they'd be dragging their wooden buoys under: salmon, trout, sometimes sturgeon. The Columbia was a great river in those days. One of the greatest in the world. You know what we've got now? Dams. A dozen of 'em between here and Puget Sound. Every last foot of this river has been dammed, and do you think I've got a chance of catching a salmon here today or any other day? Not on your life."

"Don't the dams have fish ladders?"

"Sure they do. But that doesn't mean there are any salmon left to use 'em. It's a crying shame."

From up toward the border, a fancy bass boat, its elevated fishing seats sticking up like radar devices, came racing toward us, towing a water-skier. Bill's bobber bounced up and down frantically in its wake.

"There," he said with satisfaction. "That's the most action I've had here all day."

"Not like Paul Maclean in *A River Runs Through It*?"

"Don't mention that story to me."

"You don't like it?"

"No, I do not. So I'll bet you're wondering why I fish here at all. Well, why do I? Apart from the fact that it's supposed to keep me from drinking?"

"I can't imagine."

"Of course you can imagine. I come here because to a fisherman, any fishing's better than no fishing. Look. You want to see something else that's sadly diminished from former times? Stop by the Northeast Washington Fair tonight, down in Colville. I might see you there, I've got an AA meeting at seven, and I'll slide over afterward. Just don't expect too much. The fair isn't what it once was."

"What is?"

"A good ice-cold beer," Bill Bingham said. "Hope to see you at the fair tonight."

The fairground was within walking distance of my motel on Colville's main street, and I had no trouble finding it from the reflected glow of lights in the darkening sky. The midway music quickened my step, reminding me of my grandfather, who never in his life drove by a carnival or fair without stopping for at least a few minutes to stroll the midway and see what he could see.

"What should I see?" I asked the old-timer who sold me my ticket.

"Why, see it all," he said with a chuckle. "Yes, sir. See it all and enjoy it all!"

"I'd about given up on you," a gravelly voice said close by. There in his red wool jacket, as though he'd just fallen out of the sky and was none too happy with where he'd landed, was my friend Bill Bingham, toting another quart of Mountain Dew. "I've been around the whole shooting match once already. Do you want the cook's tour?"

"You lead the way," I said, and without another word Bill stalked off ahead of me.

In the dirty riding arena in front of the big white grandstand, young equestrians from local 4-H clubs were practicing late for the horse judging contests the following morning. Bill consulted his program and read some of the horses' names aloud. "Miss

Quincy Downs, Heart's Darling, Honey Bear — imagine a horse named Honey Bear!" He took an angry pull at his Mountain Dew. "Horses ought to be named Zeke and Clyde and Dan, and let it go at that. Not cute little-girl names."

I pointed at the program. "Here's one named Bubba."

Bill snorted. "Sounds like a drunk at a line dance. Bubba's no name for a self-respecting horse."

"How about this one: Poppy?"

"Better," he said grudgingly. "Not what I'd call good. Let's go see the cattle barns. Though I can tell you right now, they aren't what they used to be. Not that anything but ice-cold-you-know-what is."

On the way to the barns, Bill told me that most of the family dairy farms in northeastern Washington had gone out of business over the past thirty years, just as they had back in northern New England, and that you couldn't have much of a dairy display without dairy farms. Still, the barns were well lighted and freshly whitewashed, and the stalls were decorated with evergreen branches and fall flowers, bright zinnias and dahlias and gladiolas, and there was a festive air, with fairgoers milling through and taking it all in contentedly.

Bill was quick to point out that the horticultural building next door wasn't what it once had been, either, so I suggested that we take a swing through the midway. We hiked back the way we'd come, drawn by the hurdy-gurdy music and the pungent scents of fair food, and I bought a fat sausage loaded with diced red tomatoes, green pepper curls, and snowy white onion slices, while Bill, solicitous of his recuperating stomach, settled for another Mountain Dew.

The midway was run by Candy Apple Amusements out of Spokane — "When you have a sweet tooth for fun, call Candy Apple" — and it was as innocent as its concessionaire's name: a couple of dozen rides and games, five or six food booths, and one tame Haunted Hall of Mirrors. "All the fun's gone out of these little North Country fairs," Bill lamented. "I can remember when you couldn't walk the length of a local midway on either side of the border in these parts and not see at least one good old-fashioned slugfest. Well, so much for tradition. It's a crying shame. There's a bar over on the main drag where they carry the Mariners games. Let's slide up there and catch an inning or two."

I looked at him.

"It's okay," he said. "That's part of my regimen. Going into a bar and drinking Mountain Dew. Come on. Baseball isn't what it used to be, either, but it's like fairs and fishing. Any baseball game is better than no baseball game."

SPORTS BAR, the dim maroon light over the tavern window said, and sure enough, the Mariners were playing the Blue Jays on a sixty-inch screen swung out on a shelf from the wall, above eye level, so we had something of the illusion of being at the movies. I ordered a hamburger and a beer, and Bill ordered yet another Mountain Dew. He stared at the screen for a moment. Then he planked his soft drink down on the table and gave an exasperated sigh.

"The only kind of baseball that's ever interested me is town team baseball, and town ball's as endangered these days as the California condor," he announced.

I perked up. Having played my own share of town baseball, I was curious to hear what Bill had to say on the subject.

"I played town team baseball in every burg I ever worked in right up into my early forties," he said. "Now, why did I stop? You tell me."

"Too old?"

"Hell, no. A man's never too old to play a little baseball. You're what, about my age? Fifty? You ought to know that. I stopped because it got to where I couldn't find a team to play on. Used to be, every border town in northern Washington and southern British Columbia fielded a club. You talk about your international baseball rivalries" — he gestured contemptuously toward the Toronto-Seattle game on television — "they don't know what an international baseball rivalry is these days. When I first broke in, we had teams from Republic, Colville, Northport, going up over the line to play Canadian teams like Cascade, Trail, Rock Creek. *Those* games were rivalries. Hundreds of fans would turn out to see them. We had donnybrooks to rival the riots between the old Brooklyn Dodgers and St. Louis Cardinals. Men up in their forties and fifties, just like you and me, duking it out all over the field, with half the goddamn town joining in. I can tell you, the local players were heroes in those days. Now, I ask you, what's caused the demise of town team baseball over the past several decades?"

"TV baseball?"

"Wrong," Bill said. "What's caused the demise of town baseball is the game of softball. Did you ever see a man hit a four-hundred-and-fifty-foot home run in softball? Did you ever see a pitcher throw a yellow-hammer curve ball with one of those punky old cow plops? You can't steal a base in softball, and can you, in your wildest imaginings, conceive of some fat-ass Budweiser-guzzling softball player that can't bend over to pick up a dribbler being anybody's hero?"

He took a long, savage pull at his Mountain Dew. "Did anyone ever call softball the great American pastime?"

"I hope not."

"I hope not too. What position did I play?"

This time I was sure of my answer. "Catcher."

"Right. I could play anywhere, but my true position was catcher. I really worked those umpires."

"I'll bet you did."

"It was baseball that got me into sports writing for local papers, and sports writing that got me my first editor's job, and when I go back to work at my paper, I'm thinking about getting into sports writing again. Sports and outdoors, a few features, book reviews. I like helping young writers and I'll keep doing that but" — patting his stomach — "no more editing for this old whore. Maybe a column once a week. I've got some things saved up for a column . . ."

His voice trailed off.

"You know," Bill said after a minute, "earlier today, up on

the Columbia, you mentioned Maclean's book. *A River Runs Through It*? I told you I didn't like it. You looked surprised."

"I was. That's a book I'd think you'd admire."

"Well, for Christ's sake, of *course* I admire it. It's beautifully written and as true as any story I ever read. I didn't say I didn't admire it. I said I didn't like it. There's a vast difference."

"So what didn't you like about it, then?"

"Well, I knew that wild young brother. Paul Maclean. The hard-drinking newspaperman. I knew him well. And I knew he was going to die in the end, and I didn't want him to. I didn't want to see Paul Maclean die."

"What did you want?"

"I wanted to see him quit gambling and go on the wagon. Make a fresh start. Be a survivor. Make something out of himself."

"Then there probably never would have been a story."

"I know that, Mr. Storywriter! Stop defending the book, you don't have to. It's a great book. Okay? Does that satisfy you?"

"I understand what you're saying," I said. "Making something out of yourself, being a survivor, is the harder thing to do."

He winked and pointed the neck of his Mountain Dew at me: right. Then he leaned his big red earnest face across the table and said, "Listen. Right this second, I want an ice-cold beer, and another, and another, more than anything else in the world. Except one thing."

"Which is?"

"Tomorrow. I want to go fishing tomorrow for bass, maybe come see those emasculated horses with cute little-girl names, even scout up a softball game to watch if I get really desperate. There isn't much up here in this North Country that's quite what it used to be and that's a goddamn crying shame. But it's better than nothing. I'm fifty-one years old, and I want to go back to my paper. I want, by God, tomorrow!"

The Inferno and the Desert

I was standing where the fire jumped the trail. At first it was no bigger than a small Indian campfire, looking more like something you could move up close to and warm your hands against than something that in a few minutes could leave your remains lying in prayer with nothing on but a belt.

— Norman Maclean, *Young Men and Fire*

5:30 A.M. This morning in the upper Columbia Valley of northern Washington dawns gray. I think it may rain but hope not. A lot of local kids are banking on a sunny day for the 4-H horse show, and so is Bill Bingham. Peering up at the sky, I fear for his spirits in bad weather, though probably I don't need to. After all, he knows better than anyone else that there'll be rainy days ahead mixed in with the sunny ones.

7:30 A.M. Today I plan to do some hiking in the Colville National Forest. Though the weather doesn't really matter much to me, I'm far from disappointed when the sun breaks through as I swing into a pulloff near the divide of the Kettle River Range and head up a faint hiking path. As I approach the top of the ridge, out of earshot of the highway below me in the pass, the clouds continue to break apart. The pointed tops of the conifers stand out sharply against the blue sky, and I'm awed by the solemn beauty of this place until, cresting the height, I suddenly look out over a wasteland in which not a live tree over two feet tall is to be seen. Except for a few blackened snags, the entire forest here has been burned to flinders for as far as the eye can see.

9:00 A.M. The site of the Colville inferno. Throughout the nineteenth century, forest fires were a fact of life that North Country natives from coast to coast, wherever there were forests, lived with every year. In 1825 upward of a million acres of Maine woodlands were destroyed by fire. In 1910 an inferno destroyed three million acres of wilderness in Montana and Idaho in just forty-eight hours. In 1912 my home state, Vermont, recorded more than a thousand forest fires, and to this day forest fires remain the most feared natural hazard and the principal cause of economic devastation in the North Country. Here in the West a select outfit of men and women have made a profession of fighting them. They're known as smokejumpers, from their practice of nonchalantly parachuting down into raging conflagrations from light Forest Service airplanes. In the elitism of their woodsmanship and courage and sheer derring-do, smokejumpers are the closest modern-day equivalents I know of to the voyageurs and river drivers of the past two centuries — a breed apart — and on the charred ridgeline of the Kettle Range this morning, I'm lucky enough to meet two of them, whom I'll call Ray and John.

In 1988, at the time of the great fire that would destroy twenty thousand acres of forest land in this area, Ray and John were college students in Spokane, putting themselves through school by smokejumping. Now living in Oregon, where Ray is a professional forester and John teaches high school biology, they make a pilgrimage to the scene of the Colville fire each fall to see how the big woods they fought to save is recovering.

"Slowly," Ray tells me. A tall, scholarly-looking young man with wire-rimmed glasses, he speaks carefully, measuring his words. "Very, very slowly. This fire started from lightning, the way the vast majority of forest fires do. Lightning struck a tree near where we're standing right now, experts think. The afternoon we were dropped in here, the wind was blowing thirty-five miles an hour and the flames were a hundred feet high. We went right to work digging a fire line over on that second ridge to the west, but by then the fire was in the treetops. That's known as a crown fire. When it got to our trench later that afternoon, it just leaped over to the trees on the other side in the air. That happened at least a dozen times over the next week."

"Hearing those crown fires racing at us was the most frightening experience of my life," John says. He's a solidly built, dark-haired man who, it turns out, played football in college and now coaches the football team at the high school where he teaches. He doesn't look like he'd be afraid of much, so when he says that the advancing crown fires were terrifying, you know he isn't exaggerating. "We lit some backfires in front of the main fire, hoping the wind would push the flames back toward the big blaze, but the wind didn't cooperate, and we wound up having to fight our own backfires. We were all scared out of our wits there'd be a blowup. That happens when the air between the backfire and main fire gets superheated and suddenly explodes into flames for acres around, like a mini–nuclear explosion. Thankfully, we never had one. But the fire burned on for weeks and, as you can see, did unbelievable damage."

"The only upside of all this is that during a big fire the lodgepole pine cones open and distribute their seeds," Ray says. "You won't recognize this place twenty years from now. It'll all be woods again. Frankly, though, what we'd both really like to do is come back here to live. We grew up locally and this is our country. It's where our hearts are."

"Can't you find jobs in the area?"

Ray and John look at each other. Then Ray says, "John and I have been together for five years, since we were in college. And the truth is, we don't think these little northern communities are the right place for us socially."

"Hell, we know they aren't," John says, laughing yet not laughing. "If we tried to come back here and settle down, we'd very probably find out what a real inferno is."

11:00 A.M. I'm descending from tree-clad mountains into an arid terrain of bare rock and brown sand and sage and cactus. This country looks more like the Mexican than the Canadian border; and when I stop in Tonasket for coffee at a diner crowded with men and women speaking Spanish, I've got to wonder what under the sun's going on. Looking out the diner window across a desert at a belt of what appears to be subtropical fruit trees gives me the jim-jams. A sojourner in a strange land must go with the flow, though, so I grab a tortilla, leave my car at the diner, and strike off toward the orchards to see if I've arrived at the Rio Grande by way of the Columbia River, or blundered onto the set of *The Milagro Beanfield War, Part II*, or maybe even done a little fancy time-traveling back to an earlier eon.

11:30 A.M. Just down the street I run into Roy Colbert, who sets me straight. An amiable, middle-aged man who manages the Chief Tonasket Apple Growers' Cooperative, Roy explains that the men and women at the diner were in fact Mexican orchard workers on their noon break and that I have in fact come to an authentic North Country desert. "We only have an eight-inch annual rainfall average here in the Tonasket Valley. The valley lies in the rain shadow of the Cascades, and nearly all the moisture coming in off the Pacific gets dumped on the big peaks west of here." Roy gestures off toward the fruit trees along the Okanogan River. "Fortunately, we've developed an innovative pressurized irrigation system that makes all this possible."

When I tell Roy Colbert that I've been visiting people who've made fresh starts in northern Montana, Idaho, and Washington, he laughs and says he's had a fair amount of experience himself in that department. "My father owned an orchard, but we weren't by any means rich. The main problem with growing apples in the Tonasket Valley back in those days was finding reliable temporary help to pick them. See that creek over there?"

He points at a brackish trickle of water seeping through some stunted cottonwoods toward the river. "Locally that's known as the Jungle because that's where tramps off the old Great Northern spur line used to jungle up. From about the time I turned fifteen my dad would send me over there in our farm truck at harvest time to recruit help from the railroad tramps. I'd literally pick up the hoboes and put 'em in the back of the truck and run them back out to the house, and we'd fill them full of coffee and bacon and eggs and toast, and then if we were lucky they'd work for a day or two. Once they got paid, no power on earth could keep them from slipping off and buying a gallon of Wild Irish Rose and jungling up again for another binge. The next day the whole recruitment process would begin all over. I can tell you, it was quite an education for a teenage kid.

"After high school I worked my way through college as a meat cutter, and eventually I came back to Tonasket and bought an orchard. But growing apples in this country is at best a touch-and-go enterprise. A few years ago I lost everything. I wound up filing for a Chapter Eleven and getting two hundred and fifty dollars for my business. Well, it happens. If you want to survive here, you learn to take life as it comes. In some ways, every day in this country is a fresh start."

1:00 P.M. Frank Western Smith, an internationally acclaimed landscape artist in his midseventies, of Osoyoos, British Columbia, knows all about fresh starts. Like many North Country natives I'd met on my trip — like me, for that matter — Frank seems to have done a little bit of everything in addition to his main line of work. Besides painting hundreds of landscapes of the desert terrain of this part of the border, he's been a wildlife and historical lecturer for the B.C. Forest Service; a ranger in Banff National Park, where he lost the use of his left arm when he was thrown from his horse and laid up in the wilderness for several days in the dead of winter; a student at the College of Art and Design in Los Angeles; and an officer for the British Columbia provincial police. He's retraced the route of the explorer David Stuart of the Hudson's Bay Pacific Fur Company, painted much of the territory along it, and helped establish what he doesn't hesitate to call "the finest small-town museum in North America," which contains many of his paintings and tableaus of the area. My favorite? A life-size diorama showing a Prohibition-era moonshine still being raided by the mounted police.

Frank Western Smith tells me that he doesn't regard himself as a member of any particular artistic school; his landscapes and murals are wonderfully detailed, highly coloristic, with the most vibrantly polished blues and oranges I've ever seen, and, like the North Country desert he paints, some of his canvases have a touch of magic realism as well. One that particularly catches my eye is *The Ghost Train*, in which the desert landscape is visible through a transparent steam train, suggesting (to me) the vanishing of the frontier era in this part of the Northwest.

I've never before seen an effect in oils quite like the transparent train; but asking Frank Western Smith how he achieved it is like asking Houdini to demystify a vanishing act. "I put it in with great pains," he says with a Mephistophelean grin. "That's all."

On the north wall of Frank's studio on the top floor of the museum is his current work in progress, a great wall-length mural, about half completed, of trains and prairies and ranches and mountains. "Maybe that'll turn out to be my masterwork," he says. "If I live to complete it, of course."

Frank signs his letters "Western" and is enormously proud of his middle name, which he conferred upon himself decades ago. "When I was living in Vancouver, there were umpteen Frank Smiths in the telephone directory. My mail kept getting misdirected; besides, I didn't like being just another Frank Smith. So I went down to the courthouse and had 'Western' legally added to my name."

"How'd you come up with the specific name 'Western'?"

"There weren't any other Frank Western Smiths in the phone book."

"And you love your western landscape?"

Again, that wonderful, knowing smile: "That too, lad. That too. Exactly the way you love your North Country."

A North Country Love Story

Like so many of the modern-day frontiersmen I'd met on my trip, he was the last of a breed: a working cowboy, a drifter who rode a fifty-by-twenty mile swath of high-altitude range in the upper Cascades from mid-June to mid-September, living alone in a log cabin with three cow dogs, two painted horses, and an unamiable pair of red pack mules for company. During the winter, after driving the cows back down to the home ranch sixty miles away, he worked as a trail guide on dude ranches in New Mexico, wrangled stock for movie studios making westerns, worked in the woods — whatever he could scout up to do.

His first name was Cash, and though he must have had a last name, he never told me what it was and I never asked. There was a kind of professional anonymity about him that I recognized immediately and respected. When he put on his cowboy hat and vest and riding boots and chaps, he was a cowboy named Cash. The first name was enough.

His cabin was at the very end of a bad mountain road west of Osoyoos, British Columbia, on the Washington side of the border, that I just barely managed to get up without four-wheel drive. From his corral you could look north toward the border and see ten snowy peaks and three hanging glaciers. An icy trout stream lazied down through the meadow past his place, but I hadn't come here primarily to fish. The home rancher, who owned the cattle and leased the range from the National Forest Service, had told me that Cash was going up into the high peaks by the border on a two-day trip to bring the last of the herd down from their summer range and that if I asked him, he'd probably let me tag along.

I arrived just after sunrise, the day after visiting Frank Western Smith, and found Cash saddling his horse and mules. He planned to stay overnight in an outpost camp for elk hunters, and the mules would be toting in camp supplies for the hunters to use later that fall. I told him who I was and that I'd love to come alone, then waited for an answer that he seemed in no hurry to give.

Quietly speaking the mules' names, Cash continued to load them with camping gear, nonperishable food, sleeping bags, a Coleman stove. "This one's kind of snuffy," he said. "Once he's packed, he's okay. But if he takes it in his head to pitch a fit and roll over on the packs while you're loading, you can't stop him for love or money."

He talked on as he worked, as much to calm his mules, I thought, as to be sociable. He told me that he'd first come up to the High Meadows cow camp ten years ago, when he was just nineteen. He'd driven up to see the cowboy who worked here then, a friend of his from Colorado. While Cash was visiting, his friend abruptly decided to light out for Alaska. The home rancher asked Cash if he could ride. "Some," he replied. That was enough for the rancher, who had told me the night before that Cash was the best hand he'd had at the High Meadows camp in thirty years.

In a typical season in that thousand square miles of wilderness range, protecting his cows from summer blizzards, lightning storms, grizzly bears, mountain lions, coyotes, golden eagles, rustlers, and their own tendency to wander deep into the trackless forest, Cash lost three or at most four animals. Today, he said, he was going after five hundred cow-calf pairs in meadow fingers up near the glaciers. "Did I ever tell you the story about the dudes I took fishing up there back five, six years ago?" he said, and was off and running, though I hadn't been there ten minutes and we'd never met before in our lives. And I didn't know what his answer to my request to go with him was until he began saddling up his second horse. All he said was, "This is a good one. He's never gotten snuffy on a rider once in his life."

It was a beautiful fall morning, clear as a bell, with the snowy mountains and the distant glaciers sparkling in the sunshine. Cash went first on his big paint, leading the two red mules behind him on a long rope. Now that we were on the move, they seemed fine. Except for a tendency to lollygag behind and put its head down to graze, then break into a jolting trot to catch up, my horse did fine, too. "He does that a-purpose," Cash said. "He likes to trot. Actually, he's Sylvie's horse. Did I ever tell you how she galloped him after a bull elk and nailed that elk dead with one shot from the saddle?"

Sylvie? I was curious but said nothing. As my brush-selling mentor, Bernie the Great, had taught me many years ago, the thing to do now was to keep still and listen to what Cash had to say, whenever he was ready to say it.

In the meantime I listened to the brook rippling down through the long natural meadows, to the easy morning breeze in the tall lodgepole pines and Douglas firs that crowded right up next to both sides of the trail, to the bawling of the red-and-white Hereford cows that Cash had brought down to the lower meadows earlier that month, to the occasional splash of a trout taking a live grasshopper that the horses scared into the brook, to the steady, soothing clop of the animals' hoofs on the packed trail. This was gorgeous country, as spectacular as any I'd seen since starting out on my trip. The grazing meadows were half a mile to a mile long and about a quarter of a mile wide, scattered at irregular intervals through the deep forest. Each time we came out of the woods into another pasture, we could see more snow peaks ahead and more glaciers.

At noon we stopped, and while I caught half a dozen trout, Cash built a neat, economical fire. He cooked the fish with bacon, and we ate them in bread-and-butter sandwiches and drank three cups of cowboy coffee apiece. Over the third cup of coffee, Cash began talking about Sylvie.

"I don't know if she's coming back or not, but I sure hope so," he said. "At least to pick up her horse, and so's I can see the little girl again."

He paused.

"Sylvie lived up here with me for four years," he continued after a minute. "She and her girl. Come south with me winters, too. She'd run away from a bad husband, you know, a mean old man old enough to be her father. He had money but he didn't treat her or the little girl good a-tall."

Cash shook his head. "That Sylvie. She could ride, rope, fish, track an elk through snow and shoot it and butcher it and fry you up the best elk steak in the world. She could fight, too, when I got snuffy. She'd had to, to keep the old man from hurting her and the girl. Once I come home from elk camp so drunk I couldn't hardly sit my horse, and Sylvie near to kilt me, she fought me so hard. Said if I ever did such a thing as that again she'd leave me. Right there was the end of my drinking days. But I reckon she got tired of being alone up here, nobody for the girl to play with, and drifting around from pillar to post in the wintertime. 'Cause when the old man showed up with his big old Cadillac car and his big pocketbook and a lot of big talk about how he was reformed and going to take good care of her and the gal, she went back with him. She let that old son of a bitch waltz her back east with nary a goodbye to me a-tall. I believe she might come back yet, though. I hope so anyway."

"I hope so, too," I said, and we headed out again.

Later, high in the mountains near the border, Cash and I unloaded the mules and hitched them to two trees. Then we went up into the meadow fingers and I watched him and his three mongrel dogs round up the last of the cows, yelling and barking and galloping hard to head off wayward calves, exactly as cowboys have done from here to Mexico for the past hundred and fifty years and more. "Hup, hip, ho!" Cash yelled. "Hup, hip, ho!" By late afternoon he'd gathered all five hundred pairs of bellowing, bawling red-and-white beef cattle and had driven them down past the elk camp into a gated meadow, where we'd pick them up on our way out in the morning.

Then a near disaster. Just before we quit for the day, in a thickly forested area, my saddle worked loose. Before I knew what had happened, I'd slipped underneath the horse. As I slid, I had the sensation of hitting black ice in my car, having no control at all. There wasn't a thing I could do. Worse yet, my left foot got caught in the stirrup, and when I jerked it free, the horse reared up and the big blue sky above the Cascades was suddenly full of bucking horse. He came down on my right leg with both front feet, and I was sure it was broken in at least ten places. I crawfished out from under the rearing horse and willed myself to jump up and take a step, then another. Nothing broken after all, but I was so light-headed from the pain and shock, I had to kneel again. My head was swimming and I was sure I was going to be sick.

"You okay?" Cash said as he calmed down the horse and tightened the saddle.

"I guess so."

"Good. You want to climb back on, then, and ride over by that big red boulder so's the cows don't head up that side finger? I'll get the gate."

That night at the outpost camp, while I nursed my bruised leg — now fast turning black and blue from ankle to groin — with aspirin and a coffee mug full of the whiskey we'd brought in for the elk hunters, Cash told me the story of Sylvie again. He felt terrible, and listening to him, so did I. He was as broken-hearted as poor Dish, the lovelorn young cowpuncher in *Lonesome Dove*. He spoke of finding something steady to do in the winter, maybe even giving up cowboying altogether. Somehow he'd find a way to take better care of Sylvie and her little girl.

From up in the fingers, near the glaciers, we heard two bull elks bugling to each other, a high, clear, piercing sound like no other I'd ever heard. And though my leg throbbed some despite the aspirin and booze, I was relieved that it wasn't broken, and oddly pleased that at fifty I could still have what amounted to a close call and survive it with relatively little to-do. But that wouldn't solve Cash's problem.

After a while the fire got low. From down the trail past the gate, a mile and more away, the half-wild range cattle had stopped lowing. It was getting late. Still, Cash talked on and on about Sylvie, hoping against hope that she'd show up before he drove the cattle back to the home ranch later that month and shut down the High Meadows cow camp for another winter.

I started to doze off. Just before I fell asleep, though, a question occurred to me, the only one I asked Cash about Sylvie: how long had she been gone?

There was a long pause. Then Cash said in a voice that was just audible, "Three years this month."

After that neither of us spoke again until morning.

Crossing the Cascades

2:00 P.M. It's raining hard in Princeton, British Columbia, and a rodeo I'd hoped to see here has been canceled. When I stop for gas downtown, the rain begins to turn to ice on my windshield. Just north of town I run into a blinding squall. First it's rain with a mixture of sleet. Soon the sleet turns to huge wet flakes of what we call sugar snow back in New England because it's apt to fall at maple sugaring time, and before I know it I'm in the middle of a swirling snowstorm, which like the prairie wind back on the Saskatchewan-Montana border seems to come at the car from all directions, cutting down visibility to just a few yards in front of my headlights. In less than two hours I've jumped from the dog days of late summer to deep winter.

3:00 P.M. The storm's blown over, the sun's out, and the sky is blue and cloudless above the snowy peaks behind me. I'm out of the Cascades and back into late summer in a countryside more closely resembling northern New England than any I've seen since Vermont. Here in British Columbia's Fraser River Valley are tall blue Harvestore silos, spanking white houses, and fire engine–red barns. Grazing pastures roll gently away toward wooded hills. And in the background, just across the border in Washington, looming more than ten thousand feet high and covered halfway down its sides with fresh snow, is Mt. Baker: a genuine North Country volcano, which erupted as recently as 1881 and, judging from the ominous clouds of steam that spouted from its cone less than twenty years ago, could erupt again at the drop of a hat.

5:00 P.M. I cross the Canadian-American border for the last time at Sumas, Washington, and for the first time I'm completely stonewalled by a customs guy, a big man of about fifty-five with a bullet-shaped head, who tells me flatly, "I have a long-standing aversion to talking to anyone in the press that dates back to Vietnam."

Okay. Fair enough. It's getting late, and at the moment all I can muster curiosity about is the confiscated mount of a polar bear shot illegally in Alaska that someone tried to sneak back across the Sumas border in a truck. The customs agent is as good as his word, though, and when I ask about the bear he just shakes his head and says, "Once burned, twice shy," and I know that I might as well ask the bear itself to tell me its history. Back in Maine I might have persisted, but my trip has mellowed me a bit toward officialdom. I've found most U.S. and Canadian customs and immigration people polite and forthcoming and, despite their uniforms and guns and the authority therein invested, helpful. Besides, I don't want my last border crossing to end on a sour note. So instead of trotting out my letters of introduction, which I'm pretty sure won't cut any ice with Major Sumas (as I think of him) anyway, I just nod and put out my hand, and after a split second's hesitation, the major grins and extends his.

"Look," he says. "I just don't want to talk. It's a personal thing with me, nothing at all to do with you, okay? But I can give you some names of local people and some tips on where to go." Which he proceeds to do, even writing down his mother-in-law's name and phone number; and just before I leave, he tells me to be sure to drive out to the commercial fishing dock in Blaine the next morning, where the land border ends at the Pacific. He walks me to my car, and the last thing he says, after shaking hands again, is, "Nothing personal, you understand. It's just that, well, you probably wouldn't want to hear what I've got to say anyway."

He's wrong about that, but I thank him for his tips and drive away wondering if maybe he really did want to talk to me about his life and if I may have missed out on a good story by not pressing a little harder. Sometimes, though, an untold story can resonate just as long — and for some reason, Major Sumas's has continued to resonate with me.

Turning Fifty in the North Country

One of the deepest satisfactions of a rural northern existence is the continuity of life from season to season, year to year, youth to old age, generation to generation. So many of the people I'd met were still living where their parents and grandparents had lived and doing the same kinds of work — farming, ranching, railroading, fishing, logging, whatever. And one of the unexpected benefits of my trip through the North Country was the opportunity it afforded me to reflect on the continuity in my own life. Maybe that's what the gypsy fortuneteller was alluding to when she told me I'd take a journey, but that it might turn out quite differently from what I expected.

The dusky countryside of coastal Washington, like the Fraser River Valley just across the border, reminded me of Vermont. As I drove, I found myself thinking how, nearly every day at home, instead of sitting down alone for lunch, I grab an apple or slap together a sandwich and go for a long tramp on some high fields partway up the mountain across the road from my house, where you can still walk ten miles due west, toward the northernmost peaks in the Green Mountain chain, and not strike a single road or building or any sign of another human being. That is where I went earlier that summer, on the June day I turned fifty. On the way back to my house in town, past a lone clump of late-blooming daffodils marking the site of a long-gone farmhouse that had overlooked one hundred miles of Vermont and Canadian mountains, I thought to myself: set me down here in this beautiful spot at any time of the year, except possibly deepest winter, and I could give you the date within a week or ten days.

In early May the flowering shadblow tells me it's time to dig two dozen garden worms and fish the brook that flows out of the beaver bog partway up the mountain. The wild cherry blossoms a week or two later signify that I can start fishing with wet flies, and remind me of my early fly-fishing trips up the brook behind our Catskill home in cherry-blossom time with my father and uncle. In early June, on a day like today, the daisies and red and orange paintbrush grow thick along the edge of the lane. I can stand in them and look out over the entire northern half of the Kingdom, across the border and deep into Canada, and it all looks exactly the same as it did years ago when Phillis and I first arrived here.

It's a small pleasure, coming here to this mountain each day. Yet it's exactly such small pleasures that invest my North Country life with the continuity that might otherwise have come from, say, beginning a new semester of teaching, if I'd chosen an academic career years ago; and now, on the day of my fiftieth birthday, I thought that for me this life in the Kingdom had been the right life, that its simple, continuous satisfactions — walking the mountain each day of the year, visiting with friends, fishing for brook trout, watching a pickup baseball game on the village green, sitting quietly in the evening with the person I've loved for thirty years and more, reading a good old or new book, writing a new story — were enough.

The morning after my last border crossing was a Sunday, and though at home I'm not a churchgoer as a rule, I accepted an invitation to attend a sunrise service with a family I met over breakfast at a café north of Bellingham. It was a simple ceremony in a tiny evangelical church at a tiny country crossroads with no nearby town. The minister, a plainspoken farmer and part-time lay preacher about my age, spoke from the heart about the trials and satisfactions of living in northern Washington. Then he spoke about immortality. His personal vision was of a border-country heaven where people would literally ranch, work in the woods, hunt and fish, and run their own affairs, much as he and his congregation strived to do here on earth.

Twenty years ago — ten years ago — I'd have been skeptical. Now I wasn't so sure. At fifty I don't dwell on metaphysical matters as much as I used to, but as my faith in this life has increased, so too has my faith in the overall design of things. Look at the felicitous surprises this six-week trip held in store for me, not to mention the surprises of my entire life in the North Country. I couldn't have begun to predict the bountifulness of either. How then could I rule out anything about our ultimate destination? Who's to say that Sharon Butula didn't have that reassuring encounter with Wallace Stegner's spirit? Or that Stegner's ultimate journey, or that of the Washington preacher's, should be any less surprising and hopeful than a Vermont writer's midlife ramble across America's northern frontier?

Notes from the End of the Line

10:00 A.M. Overlooking the Pacific Ocean.

"There's one! Up by the mouth of that little crick. No, further over. Cast further over. Yes! Right there."

The young man beside me on this one-lane wooden bridge within sight of the Pacific could have been Bob Bagnall, back at the Soo, or my neighbor's son down the road in Vermont, or my own son for that matter. He's spotting fish for his partner, who's standing up to the tops of his waders in the pool below and casting for the big king salmon that have come up this estuary from the ocean to spawn. The tide's out, and with the guidance of his friend on the bridge, the fisherman makes a beautiful cast, less than a foot away from the cruising salmon. I know he's got his heart set on this particular fish, a gigantic thirty-pounder, which swims unperturbed and uninterested past the small gold spinner twirling and flashing just inches from its long under-shot jaw.

I've come across these salmon fishermen during the course of an after-church hike along the coast just south of Blaine and the Canadian border, on the northwesternmost point of land in the contiguous United States. It seems propitious to me this crisp blue fall Sunday morning, with the sunshine gleaming off Mt. Baker's snowy head, that the last day of my trip should have begun with a homespun vision of a North Country heaven and then immediately proceed to fishing. All across the border, from Maine to the Pacific, I found fishing to be the North Country's universal participant sport. And even though these two anglers don't really believe they can entice the big kings into striking today, they can always hope, and that hopefulness seems emblematic of the spirit of the North Country.

* * *

10:30 A.M. So I'm standing on Blaine's public dock just south of the border and looking west past the fish processing plant to the Pacific beyond, thinking that this is it — the end of the line. Now I'm going to follow Major Sumas's advice and walk down to the very end of that dock and stand on the last dry square foot of U.S. border territory, if only just to say to myself that I've done it.

Already the morning is quite warm, with something of the rarefied feel of Indian summer in the air, a time of year I've always regarded as lucky, a break for everyone before the relentless northern winter sets in in earnest. And here in the September sunshine, on the edge of the Pacific, my own long run of luck holds up when I discover Sami Munian and his teenage son, Kevin, out catching rock crabs for the family's Sunday dinner.

The Munians, who live just across the bay in Canada, are using wire traps baited with dead fish, which they fling off the end of the dock on orange nylon rope, the same kind of rope Bob Bagnall used to lower his homemade bicycle-rim net back at the Soo. Sami pulls one of the traps up out of the water to show me how it works. Inside are two undersized male crabs and several females, all of which he tosses back, and one keepable male, somewhat larger than the palm of my hand, which he flips into a bucket with some others to take home for dinner.

"I came here from the Fiji Islands nineteen years ago with nothing," Sami tells me. "Not a penny in my pocket, just my airplane ticket. Now I have a wonderful wife, four great kids, a house, and next month I'm buying a salmon trawler. Then I will become full-time captain of my own fishing boat and go out there." He points beyond Blaine's bay toward the ocean, as flat this morning as a Vermont millpond on a sunny summer day.

"With my sons," Sami says. "They'll work for me and I'll help pay for their education."

"It sounds like a dream come true."

"It is. Our very own business. We'll fish for salmon, rockfish, cod, whatever. I know it won't be easy. But I understand how to fish. In the Fijis, everyone grows up fishing. And when you get that full bag, after maybe going days without a fish, so tired and discouraged from searching for them you wouldn't believe it? Well, there's no other feeling like it. I want my boys to know that feeling. So when I take out the trawler for the first time? That will be the fourth biggest moment in my life."

"What are the other three?"

"Coming to North America. Marrying my wife. Having the boys. And now, the fishing boat. After nearly twenty years, a dream come true."

I smile and shake my head. I cannot, it seems, get away from dreams.

"You seem very happy with your life here," I say a minute later. "What's your secret? Hard work?"

Sami Munian rebaits his crab trap and throws it back into the ocean he crossed twenty years ago to make a new life for himself, the ocean he'll soon be making his living from. He reflects for a moment as the trap sinks out of sight into the oily dockside water. Then he says, "Yes, I believe that hard work is always the secret to success. But happiness? Maybe this. My guys?" He gives his son a quick pat on the back, and Kevin, half-embarrassed, pretends to duck away. "I spend a lot of time with them," Sami says. "When I'm spending time with my guys, I'm happy. That's when everything about my life comes together for me."

And at exactly this moment, at the end of the line on a dock jutting into the Pacific Ocean, visiting with a Fiji Islander who has come to this remote corner of North America to make a fresh start and now, two decades later, is about to make another, everything about my trip comes together for me, too.

Now it's time to go home and write about it.